CENTRAL STATISTICAL OFFICE

KEY DATA

1991 / 92 EDITION

London: HMSO

Key Data

INTRODUCTION

The aim of Key Data is to present you with basic statistics in the main economic and social areas. The tables selected for reproduction have almost without exception been taken direct from other publications, most of which are also produced by the Central Statistical Office:

Annual Abstract of Statistics
Regional Trends
Social Trends
Monthly Digest of Statistics
Economic Trends
Financial Statistics
United Kingdom National Accounts (The ''Blue Book'')
United Kingdom Balance of Payments (The ''Pink Book'')

The original publication is named under each reproduced table or chart, whilst within the table or chart there is normally a reference to table or chart source of the figures.

With this variety of origins it is inevitable that the tables do not always conform to a standard presentation. Nevertheless they have in common that they relate to the United Kingdom in total unless clearly stated otherwise, that figures for less than a year are seasonally adjusted unless stated otherwise, and that where symbols are used they have the following meanings:

—	*nil, or less than half the final digit*
. .	*not available*
italics	*normally indicate percentages*
†	*new or earliest revised data*
*	*average (or total) of five weeks*

CSO Databank

Some tables in this publication contain data which is available on datasets in the CSO Databank. The appropriate four digit identifier is included at the top of the column or start of the row of figures. This is to facilitate access to the data in computer-readable form and make available longer runs of data than appear in these tables.

The CSO Databank is a collection of mostly macro-economic time-series available on magnetic tape or disk. The tape format, unlabelled EBCDIC, is the same for all the datasets. The disks, either 3½" or 5¼" are written in ASCII text which can be loaded as spreadsheets and viewed using standard spreadsheet packages, such as LOTUS or SMART. Details of the service offered, and the schedule of charges, are available from the Databank Manager, Room 52A/4, Central Statistical Office, Government Offices, Great George Street, London SW1P 3AQ (Tel: 071-270 6386, 6387 or 6381).

The whole text of the latest 'Government Statistics — a Brief Guide to Sources' is reproduced on pp 89–110. This provides a useful list of the most important government publications containing statistics and sets out departmental responsibilities and contact points.

Central Statistical Office
Great George Street
London, SW1P 3AQ.
August 1991

CONTENTS

GEOGRAPHY

Definitions and sources

Standard Regions are the 11 Economic Planning Regions of the United Kingdom. They are also classified as **level I** regions for the purposes of the European Community.

Great Britain includes England, Wales and Scotland. The United Kingdom includes Northern Ireland.

For other sources see:

Guide to Official Statistics, 1990 edition (200 pages approximately fully indexed) HMSO

1.1 STANDARD REGIONS AND COUNTIES OF ENGLAND AND WALES

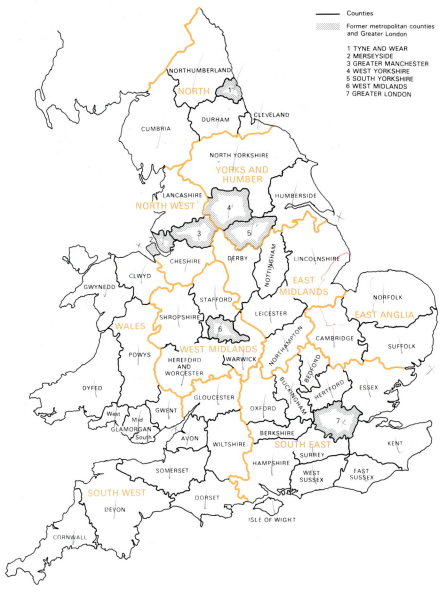

——— Counties

▒ Former metropolitan counties and Greater London

1 TYNE AND WEAR
2 MERSEYSIDE
3 GREATER MANCHESTER
4 WEST YORKSHIRE
5 SOUTH YORKSHIRE
6 WEST MIDLANDS
7 GREATER LONDON

From Regional Trends, 1991

1.2 LOCAL AUTHORITY REGIONS OF SCOTLAND
AND BOARDS OF NORTHERN IRELAND

SHETLAND
ISLANDS

ORKNEY
ISLANDS

WESTERN
ISLES

HIGHLAND

SCOTLAND

GRAMPIAN

TAYSIDE

FIFE

CENTRAL

CENTRAL
CLYDESIDE
CONURBATION

LOTHIAN

STRATHCLYDE

BORDERS

DUMFRIES AND
GALLOWAY

NORTHERN
IRELAND

NORTH
EASTERN

BELFAST

WESTERN

SOUTHERN

SOUTH
EASTERN

¹ Education and Library Boards. For Health and Social Services Boards and travel-to-work areas see Appendix.

From: Regional Trends, 1991

GEOGRAPHY

1.3 EUROPEAN COMMUNITY REGIONS

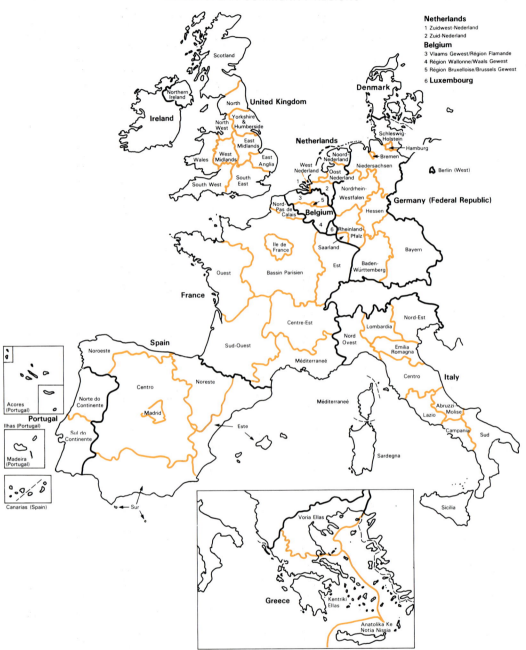

Netherlands
1 Zuidwest-Nederland
2 Zuid-Nederland
Belgium
3 Vlaams Gewest/Région Flamande
4 Région Wallonne/Waals Gewest
5 Région Bruxelloise/Brussels Gewest
6 **Luxembourg**

From: Regional Trends, 1991

Key Data 91, © Crown copyright 1991

GEOGRAPHY

1.4 EUROPEAN COMMUNITIES COMPARISONS

	Population[1] (thousands) 1988	Density (persons per sq. km.) 1988	Percentage of population Under 15 yrs 1987	65 and over 1987	Births (per 1,000 pop.) 1988	Deaths (per 1,000 pop.) 1988	Infant mortality (per 1,000 births) 1987	Dependency rate[1,2] 1989
EUR 12	324,646.5	144.0	19.0	13.9	12.0	9.9	9.0	1.2
Belgium	9,901.7	324.5	18.4	14.2	12.0	10.6	9.7	1.5
Vlaams gewest/Region flamande	5,709.2	422.5	18.3	13.5	11.5	9.7	9.4	1.5
Region bruxelloise/Brussels gewest	970.4	6,012.5	17.3	17.5	12.9	12.1	10.2	1.7
Region wallonne/Waals gewest	3,222.0	191.3	18.7	14.5	12.7	11.6	10.0	1.6
Denmark	5,129.5	119.1	17.9	15.3	11.5	11.5	8.3	0.8
France	55,883.7	102.7	20.8	13.3	13.8	9.3	7.8	1.2
Ile de France	10,370.4	863.3	20.4	10.7	16.1	7.6	7.8	1.0
Bassin-parisien	10,204.1	70.1	21.8	13.4	13.7	9.6	8.0	1.3
Nord-Pas-de-Calais	3,936.5	317.1	24.4	11.1	16.0	9.5	8.5	1.6
Est	5,040.2	104.9	21.5	11.7	13.9	8.9	8.0	1.3
Ouest	7,443.4	87.5	21.6	14.1	12.7	9.5	7.6	1.2
Sud-Ouest	5,852.6	56.5	18.0	16.9	11.2	10.9	8.3	1.3
Centre-Est	6,556.4	94.1	20.8	13.0	13.4	9.1	7.1	1.2
Mediterranee	6,479.9	96.1	18.7	16.2	12.6	10.6	7.0	1.4
Germany	61,444.7	247.1	14.7	15.2	11.0	11.2	8.3	1.1
Baden-Wurttemberg	9,377.8	262.3	15.4	14.2	11.8	9.9	7.1	1.0
Bayern	10,989.6	155.8	15.3	15.0	11.5	10.8	7.6	1.0
Berlin (West)	2,047.1	4,263.6	13.6	18.5	10.2	14.7	11.4	0.9
Brandenburg	..	..	..	..	..	..	..	..
Bremen	660.3	1,633.6	12.8	17.7	9.7	13.2	9.7	1.2
Hamburg	1,579.3	2,116.5	11.8	18.2	9.6	13.3	8.3	1.0
Hessen	5,544.6	262.6	14.3	15.3	10.4	11.2	6.8	1.1
Mecklenburg-Vorpommern	..	..	..	..	..	..	..	..
Niedersachsen	7,171.5	151.2	15.0	15.6	10.6	11.6	8.1	1.1
Nordrhein-Westfalen	16,800.8	493.2	14.6	14.8	11.1	11.1	9.4	1.2
Rheinland-Pfalz	3,642.3	183.5	15.0	15.4	10.9	11.5	9.1	1.1
Saarland	1,053.6	410.1	14.2	14.7	10.2	11.8	10.7	1.3
Sachsen	..	..	..	..	..	..	..	..
Sachsen-Anhalt	..	..	..	..	..	..	..	..
Schleswig-Holstein	2,559.7	162.8	14.2	15.7	10.7	11.9	7.2	1.1
Thueringen	..	..	..	..	..	..	..	..
Greece	10,004.4	75.8	20.3	13.5	10.7	9.2	11.7	1.5
Voreia Ellada	3,230.7	56.9	..	..	10.9	9.1	10.8	1.3
Kentriki Ellada	2,294.3	42.6	..	..	10.3	10.4	..	1.3
Attiki	3,530.1	927.0	..	..	10.5	8.3	..	1.6
Nisia	949.4	54.4	..	..	12.0	10.4	..	1.5
Ireland	3,538.0	51.4	28.7	10.9	15.3	8.9	7.9	1.6
Italy	57,451.9	190.7	18.4	13.4	10.1	9.4	9.6	1.4
Nord-Ovest	6,230.0	182.8	14.4	16.7	7.3	11.7	8.3	1.3
Lombardia	8,892.7	372.7	16.6	12.9	8.6	9.4	7.9	1.3
Nord Est	6,469.1	162.4	16.8	13.8	8.8	9.8	7.1	1.3
Emilia-Romagna	3,922.7	177.3	13.9	16.9	6.7	11.0	9.2	1.2
Centro	5,814.6	141.3	15.2	16.6	7.8	10.7	8.9	1.3
Lazio	5,146.7	299.2	18.1	12.0	9.8	8.4	10.6	1.4
Campania	5,752.2	423.1	24.2	9.8	14.7	7.8	11.4	1.6
Abruzzi-Molise	1,595.3	104.7	18.5	14.6	10.2	9.5	8.8	1.4
Sud	6,822.3	153.6	22.9	11.2	13.3	7.7	10.5	1.7
Sicilia	5,152.8	200.4	22.2	11.9	13.4	8.6	11.5	1.8
Sardegna	1,653.5	68.6	21.7	11.0	10.2	7.6	6.9	1.6
Luxembourg	373.3	144.3	16.9	13.3	12.3	10.3	9.4	1.4
Netherlands[3]	14,758.6	352.5	18.8	12.3	12.6	8.4	6.5	1.2
Noord-Nederland	1,593.0	144.9	19.5	13.4	11.9	9.2	6.2	1.4
Oost-Nederland	2,999.8	280.9	20.2	11.7	13.2	8.1	6.6	1.3
West-Nederland	6,903.9	602.6	18.1	13.2	12.8	8.8	6.1	1.1
Zuid-Nederland	3,262.0	447.4	18.7	10.5	12.2	7.5	7.5	1.2
Portugal	10,286.2	111.8	22.7	12.4	11.9	9.5	14.2	1.1
Spain	38,809.0	76.9	22.3	12.3	10.7	8.2	9.1	1.6
Noroeste	4,472.8	98.7	20.3	14.4	8.4	9.3	10.7	1.5
Noreste	4,125.6	58.6	20.2	12.6	8.8	8.2	9.9	1.6
Madrid	4,842.3	605.7	23.1	10.6	10.8	6.8	8.6	1.8
Centro	5,468.8	25.4	20.3	14.9	9.9	8.8	8.1	1.7
Este	10,444.1	173.3	21.8	12.4	10.4	8.5	8.3	1.5
Sur	7,989.5	81.0	25.5	10.6	13.5	7.9	9.8	1.8
Canarias	1,465.7	202.4	25.9	8.8	13.2	7.1	8.1	1.6
United Kingdom	57,065.6	233.8	18.9	15.5	13.8	11.3	9.3	1.0
North	3,071.0	199.4	18.9	15.3	13.1	12.3	8.7	1.0
Yorkshire & Humberside	4,912.8	318.6	18.9	15.6	13.7	11.7	9.9	1.0
East Midlands	3,970.3	254.0	18.9	15.0	13.4	11.0	9.3	0.9
East Anglia	2,034.5	161.8	18.9	16.4	13.1	10.9	7.8	0.9
South East	17,344.0	637.1	18.3	15.5	14.2	10.6	9.0	0.9
South West	4,633.9	194.3	17.6	18.1	12.6	11.9	8.5	0.9
West Midlands	5,206.5	400.1	19.3	14.5	14.0	11.0	9.5	1.0
North West	6,363.5	868.0	19.3	15.4	14.2	12.2	9.1	1.0
Wales	2,857.0	137.6	18.9	16.4	13.6	11.9	9.5	1.1
Scotland	5,094.0	64.7	18.9	14.7	13.0	12.2	..	1.0
Northern Ireland	1,578.1	111.8	25.2	12.1	17.6	10.1	..	1.3

1 Definitions of position, employment and unemployment differ from those used in UK tables.
2 Labour force sample survey, 1989.
3 Including 'centraal persoons register'.

From: Regional Trends, 1991, Table 14.1

POPULATION AND VITAL STATISTICS 2

Definitions and sources

The estimates of the resident population are based on figures from the latest census updated to allow for subsequent births, deaths and migration into and out of the country. They include residents temporarily outside the country and exclude foreign visitors. Members of HM and foreign armed forces stationed here are included, but members of HM armed forces stationed abroad are not.

More detailed statistics and commentaries are published in the OPCS Monitor series and quarterly publication *Population Trends*.

For other sources see:

Guide to Official Statistics, 1990 edition (200 pages approximately fully indexed) HMSO.

2.1 Mid-year estimates of resident population
Thousands

	England and Wales			Scotland			Northern Ireland			United Kingdom		
	Males	Females	Persons	Males	Females	Persons	Males	Females	Persons	Males	Females	Persons
	BBAE	BBAF	BBAD	BBAH	BBAI	BBAG	BBAK	BBAL	BBAJ	BBAB	BBAC	DYAY
1974	24 075	25 393	49 468	2 519	2 722	5 241	755	772	1 527	27 349	28 887	56 236
1975	24 091	25 378	49 470	2 516	2 716	5 232	753	770	1 524	27 361	28 865	56 226
1976	24 089	25 370	49 459	2 517	2 716	5 233	754	769	1 524	27 360	28 856	56 216
1977	24 076	25 364	49 440	2 515	2 711	5 226	754	769	1 523	27 345	28 845	56 190
1978	24 067	25 375	49 443	2 509	2 704	5 212	754	770	1 523	27 330	28 848	56 178
1979	24 113	25 395	49 508	2 505	2 699	5 204	755	773	1 528	27 373	28 867	56 240
1980	24 156	25 448	49 603	2 501	2 693	5 194	755	778	1 533	27 411	28 919	56 330
1981	24 160	25 474	49 634	2 495	2 685	5 180	754	783	1 538	27 409	28 943	56 352
1982	24 143	25 459	49 601	2 489	2 677	5 167	754	784	1 538	27 386	28 920	56 306
1983	24 176	25 478	49 654	2 485	2 665	5 150	756	787	1 543	27 417	28 931	56 347
1984	24 244	25 519	49 764	2 484	2 662	5 146	760	791	1 550	27 487	28 972	56 460
1985	24 330	25 594	49 924	2 480	2 656	5 137	763	795	1 558	27 574	29 044	56 618
1986	24 403	25 672	50 075	2 475	2 646	5 121	768	798	1 567	27 647	29 116	56 763
1987	24 493	25 750	50 243	2 471	2 641	5 112	773	802	1 575	27 737	29 193	56 930
1988	24 576	25 817	50 393	2 462	2 632	5 094	774	804	1 578	27 813	29 253	57 065
1989	24 669	25 893	50 562	2 460	2 630	5 091	777	806	1 583	27 907	29 330	57 236
1990	24 766	25 953	50 719	2 467	2 636	5 102	780	809	1 589	28 013	29 398	57 411

Figures may not add due to rounding.

Sources: Office of Population Censuses and Surveys; General Register Office (Scotland); General Register Office (Northern Ireland)

2.2 Age distribution of estimated resident population at 30 June 1989
Thousands

	Resident population										
	England and Wales		Wales		Scotland		Northern Ireland[1]		United Kingdom		
	Males	Females	Males	Females	Males	Females	Males	Females	Males	Females	Persons
0-4	1 713.8	1 632.4	97.2	92.5	166.7	158.6	69.5	65.7	1 950.7	1 857.0	3 807.7
5-9	1 629.4	1 548.4	93.8	89.0	164.7	157.1	68.2	65.2	1 862.0	1 770.7	3 632.7
10-14	1 517.1	1 434.1	88.6	83.3	158.1	149.7	64.8	62.4	1 739.3	1 645.2	3 384.5
15-19	1 830.1	1 739.0	105.1	100.6	190.1	183.5	70.0	63.5	2 091.2	1 987.8	4 079.0
20-24	2 077.5	2 010.1	113.0	110.6	217.8	208.4	71.8	63.8	2 367.9	2 282.8	4 650.7
25-29	2 058.9	2 023.2	113.1	114.2	210.8	201.8	64.8	60.6	2 333.5	2 285.1	4 618.7
30-34	1 761.4	1 737.0	93.4	91.2	184.2	182.1	54.3	55.3	1 998.2	1 972.9	3 971.1
35-39	1 682.8	1 681.4	92.2	92.6	167.8	168.1	47.5	48.7	1 898.1	1 897.4	3 795.6
40-44	1 821.8	1 812.7	98.8	99.6	172.6	172.8	47.0	47.9	2 041.1	2 033.4	4 074.5
45-49	1 466.6	1 454.9	83.0	82.8	144.0	149.1	43.1	44.6	1 652.6	1 647.2	3 299.9
50-54	1 367.3	1 367.4	77.1	77.6	136.9	145.7	37.7	39.4	1 541.5	1 552.2	3 093.7
55-59	1 293.7	1 323.7	74.1	77.3	131.2	144.3	34.0	37.2	1 458.8	1 505.4	2 964.2
60-64	1 245.2	1 333.4	74.2	81.5	123.1	141.1	31.3	36.5	1 399.7	1 510.8	2 910.4
65-69	1 215.0	1 422.9	73.3	86.2	113.2	142.0	28.0	35.1	1 356.6	1 600.6	2 957.1
70-74	784.9	1 050.7	48.3	65.4	73.2	106.3	20.6	29.2	878.4	1 185.9	2 064.2
75-79	648.3	1 022.9	37.1	60.2	59.1	99.6	15.5	25.4	723.1	1 147.8	1 870.9
80-84	365.7	734.1	20.5	43.0	31.9	70.1	8.5	17.3	405.9	820.9	1 226.8
85 and over	189.3	565.2	10.6	31.9	15.0	50.3	3.8	11.3	208.1	626.6	834.7
0-14	4 860.4	4 614.8	279.6	264.7	489.5	465.3	202.5	193.3	5 552.0	5 272.9	10 824.9
15-64	16 605.4	16 482.8	924.1	928.1	1 678.4	1 696.8	501.5	497.4	18 782.6	18 675.0	37 457.6
65 and over	3 203.3	4 795.8	189.8	286.7	292.4	468.2	76.4	118.3	3 572.0	5 381.8	8 953.8
All ages	24 669.1	25 893.4	1 393.5	1 479.6	2 460.4	2 630.3	780.4	809.0	27 906.5	29 329.7	57 236.2

Figures may not add due to rounding.
1 For Northern Ireland: population at June 1990.

Sources: Office of Population Censuses and Surveys; General Register Office (Scotland); General Register Office (Northern Ireland)

From: Monthly Digest of Statistics, June 1991, Tables 2.1 and 2.2

POPULATION AND VITAL STATISTICS

2.3 Population: by selected age bands
United Kingdom

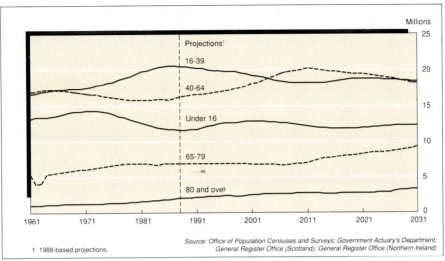

Projections[1]

16-39

40-64

Under 16

65-79

80 and over

Millions

Source: Office of Population Censuses and Surveys; Government Actuary's Department;
General Register Office (Scotland); General Register Office (Northern Ireland)

1 1988-based projections.

From: Social Trends, 1991, Chart 1.1

2.4 Age and sex structure of the population[1]

United Kingdom   *Millions*

	Under 16	16 – 39	40 – 64	65 – 79	80 and over	All ages
Mid-year estimates						
1951	. .	. .	15.9	4.8	0.7	50.3
1961	13.1	16.6	16.9	5.2	1.0	52.8
1971	14.3	17.5	16.7	6.1	1.3	55.9
1981	12.5	19.7	15.7	6.9	1.6	56.4
1986	11.7	20.6	15.8	6.8	1.8	56.8
1989	11.5	20.4	16.3	6.9	2.1	57.2
Males	5.9	10.3	8.1	3.0	0.6	27.9
Females	5.6	10.1	8.2	3.9	1.4	29.3
Mid-year projections[2]						
1991	11.7	20.2	16.5	6.9	2.2	57.5
1996	12.5	19.8	17.0	6.8	2.4	58.5
2001	12.8	19.2	18.0	6.7	2.5	59.2
2006	12.6	18.4	19.4	6.6	2.6	59.6
2011	12.1	18.1	20.2	7.0	2.7	60.0
2025	12.1	18.6	19.0	8.5	2.9	61.1
Males	6.2	9.5	9.5	3.9	1.1	30.2
Females	5.9	9.1	9.5	4.6	1.8	30.9

1 See Appendix, Part 1: Population and population projections.
2 1988-based projections.

Source: Office of Population Censuses and Surveys;
Government Actuary's Department;
General Register Office (Scotland);
General Register Office (Northern Ireland)

From Social Trends 1991, Table 1.2

2.5 Conceptions: by age of woman, marital status and outcome

England & Wales   Percentages and thousands

	1971	1976	1981	1986	1987	1988
Conceptions to women aged under 20						
Inside marriage						
Maternities	33.2	29.8	23.2	13.1	11.7	10.5
Legal abortions[1]	0.6	1.1	0.9	0.6	0.6	0.5
Outside marriage						
Maternities inside marriage	30.0	20.0	15.4	9.5	8.2	7.2
Maternities outside marriage[2]						
-- joint registration	5.8	8.7	14.8	26.6	28.5	29.9
-- sole registration	13.5	13.4	14.8	17.4	17.0	16.7
Legal abortions[1]	16.9	27.0	31.0	32.8	34.0	35.3
All conceptions (= 100%) (thousands)	133.1	105.7	115.2	118.8	123.2	120.7
All conceptions						
Inside marriage						
Maternities	72.6	70.7	65.9	58.1	55.8	54.1
Legal abortions[1]	5.2	6.1	5.6	4.7	4.6	4.5
Outside marriage						
Maternities inside marriage	8.1	5.8	5.5	5.1	4.9	4.7
Maternities outside marriage[2]						
-- joint registration	3.5	4.2	6.8	12.6	14.1	15.2
-- sole registration	4.1	3.8	4.8	6.1	6.3	6.3
Legal abortions[1]	6.7	9.4	11.4	13.4	14.4	15.2
All conceptions (= 100%) (thousands)	835.5	671.6	752.3	818.9	850.4	849.5

1 Legal terminations under the *1967 Abortion Act.*

2 Births outside marriage can be registered by the mother only (sole registrations) or by both parents (joint registrations).

Source: Office of Population Censuses and Surveys

From Social Trends 1991 Table 2.26

2.6 Births

Annual averages or calendar years

	Live births			Sex ratio	Rates[1]				
	Total	Male	Female		Crude birth rate[2]	General fertility rate[3]	TPFR[4]	Still-births	Still-birth rate
United Kingdom									
1900 – 02	1 095	558	537	1 037	28.6	115.1	..	..	..
1910 – 12	1 037	528	508	1 039	24.6	99.4	..	..	..
1920 – 22	1 018	522	496	1 052	23.1	93.0	..	..	..
1930 – 32	750	383	367	1 046	16.3	66.5	..	..	..
1940 – 42	723	372	351	1 062	15.0	..	1.89	26	..
1950 – 52	803	413	390	1 061	16.0	73.7	2.21	18	..
1960 – 62	946	487	459	1 063	17.9	90.3	2.80	18	..
1970 – 72	880	453	427	1 064	15.8	82.5	2.36	12	13
1980 – 82	735	377	358	1 053	13.0	62.5	1.83	5	7
1978	687	353	334	1 059	12.2	60.8	1.76	6	8
1979	735	378	356	1 061	13.1	64.1	1.86	6	8
1980	754	386	368	1 050	13.4	64.9	1.89	6	7
1981	731	375	356	1 053	13.0	62.1	1.81	5	7
1982	719	369	350	1 054	12.8	60.6	1.78	5	6
1983	721	371	351	1 058	12.8	60.2	1.77	4	6
1984	730	373	356	1 049	12.9	60.3	1.77	4	6
1985	751	385	366	1 053	13.3	61.4	1.80	4	6
1986	755	387	368	1 053	13.3	61.1	1.78	4	5
1987	776	398	378	1 053	13.6	62.3	1.82	4	5
1988	788	403	384	1 049	13.8	63.2	1.84	4	5
1989[5]	777	398	379	1 051	13.6	62.4	1.81	4	5

1 Rates are based on a new series of population estimates which use a new definition and population base taking into account the 1981 Census results.
2 Rate per 1 000 population
3 Rate per 1 000 women aged 15–44.
4 Total period fertility rate is the average number of children which would be born per woman if women experienced the age-specific fertility rates of the period in question throughout their child-bearing life span. UK figures for years 1970–72 and earlier are estimates.

5 Provisional.

Sources   Office of Population Censuses and Surveys; General Register Office (Scotland); General Register Office (Northern Ireland)

From: Annual Abstract of Statistics, 1991, Table 2.16

POPULATION AND VITAL STATISTICS

2.7 Live births: by country of birth of mother
Great Britain   Thousands and percentages

| | Live births (thousands) | | | |
	1971	1981	1988	1989
Area/country of birth of mother				
United Kingdom	777.3	617.3	675.5	667.7
Percentage of all live births	*88.9*	*87.7*	*88.9*	*88.9*
Outside United Kingdom				
Irish Republic	22.5	8.6	6.7	6.8
Old Commonwealth	2.7	2.6	2.9	3.0
New Commonwealth and Pakistan				
India	13.7	12.6	9.7	9.0
Pakistan and Bangladesh	8.5	17.0	17.8	17.9
Caribbean	12.6	6.3	4.2	4.1
East Africa	2.2	6.7	7.1	6.9
Rest of Africa	3.0	3.6	4.3	4.8
Other New Commonwealth	6.2	8.4	8.8	8.3
Total New Commonwealth and Pakistan	46.2	54.6	52.0	50.9
Other European Community	20.4	6.1	7.6	7.8
Rest of the world		14.1	14.9	15.0
Total with mother born outside United Kingdom	91.8	86.0	84.2	83.4
Not stated	4.8	0.2	0.1	0.1
Total live births	869.9	703.5	759.8	751.2

Source: Office of Population Censuses and Surveys;
General Register Office (Scotland)

From: Social Trends, 1991, Table 1.11

2.8 Marriages: by type
United Kingdom   Thousands and percentages

	1961	1971	1976	1981	1985	1986	1987	1988	1989
Marriages (thousands)									
First marriage for both partners	340	369	282	263	257	254	260	253	252
First marriage for one partner only									
Bachelor/divorced woman	11	21	30	32	32	34	34	34	35
Bachelor/widow	5	4	4	3	2	2	2	2	2
Spinster/divorced man	12	24	32	36	38	38	39	39	38
Spinster/widower	8	5	4	3	2	2	2	2	2
Second (or subsequent) marriage for both partners									
Both divorced	5	17	34	44	47	48	47	50	50
Both widowed	10	10	10	7	6	6	5	5	5
Divorced man/widow	3	4	5	5	4	4	4	4	4
Divorced woman/widower	3	5	5	5	5	5	5	5	5
Total marriages	397	459	406	398	393	394	398	394	392
Remarriages[1] as a percentage of all marriages	*14*	*20*	*31*	*34*	*35*	*35*	*35*	*36*	*36*
Remarriages[1] of those divorced as a percentage of all marriages	*9*	*15*	*26*	*31*	*32*	*33*	*32*	*33*	*34*

1 Remarriage for one or both partners.

Source: Office of Population Censuses and Surveys;
General Register Office (Scotland)

From: Social Trends, 1991, Table 2.10

POPULATION AND VITAL STATISTICS

2.9 Marriage and divorce: EC comparison
1981 and 1988

Rates

	Marriages per 1,000 eligible population		Divorces per 1,000 existing marriages	
	1981	1988	1981	1988
United Kingdom	7.1	6.9	11.9	12.3
Belgium	6.5	6.0	6.1	8.4
Denmark	5.0	6.3	12.1	13.1
France	5.8	4.9	6.8	8.4
Germany (Fed. Rep.)	5.8	6.5	7.2	..
Greece	7.3	4.8	2.5	..
Irish Republic	6.0	5.1	0.0	0.0
Italy	5.6	5.5	0.9	2.1
Luxembourg	5.5	5.5	5.9	
Netherlands	6.0	6.0	8.3	8.1
Portugal	7.7	6.9	2.8	..
Spain	5.4	5.5	1.1	..

Source: Statistical Office of the European Communities

From: Social Trends, 1991, Table 2.12

2.10 Divorce[1]

	1961	1971	1976	1981	1984	1985	1986	1987	1988	1989
Petitions filed[2] (thousands)										
England & Wales										
By husband	14	44	43	47	49	52	50	50	49	50
By wife	18	67	101	123	131	139	131	133	134	135
Total	32	111	145	170	180	191	180	183	183	185
Decrees nisi granted (thousands)										
England & Wales	27	89	132	148	148	162	153	150	155	152
Decrees absolute granted (thousands)										
England & Wales	25	74	127	146	145	160	154	151	153	151
Scotland	2	5	9	10	12	13	13	12	11	12
Northern Ireland	—	—	1	1	2	2	2	2	2	2
United Kingdom	27	80	136	157	158	175	168	165	166	164
Persons divorcing per thousand married people										
England & Wales	2.1	6.0	10.1	11.9	12.0	13.4	12.9	12.7	12.8	12.7
Percentage of divorces where one or both partners had been divorced in an immediately previous marriage										
England & Wales	*9.3*	*8.8*	*11.6*	*17.1*	*21.0*	*23.0*	*23.2*	*23.5*	*24.0*	*24.7*
Estimated numbers of divorced people who had not remarried (thousands)										
Great Britain										
Men	101	200	405	653	847	918	990	1,047	..	..
Women	184	317	564	890	1,105	1,178	1,258	1,327	..	..
Total	285	517	969	1,543	1,952	2,096	2,248	2,374	..	..

1 This table includes annulment throughout. See Appendix, Part 2: Divorce.
2 Estimates based on 100 per cent of petitions at the Principal Registry together with a 2 month sample of county court petitions (March and September).

Source: Office of Population Censuses and Surveys;
Lord Chancellor's Department;
General Register Office (Scotland)

From: Social Trends 1991 Table 2.13

POPULATION AND VITAL STATISTICS

2.11 Percentage of women cohabiting: by age

Great Britain   Percentage and numbers

	1979	1981	1986	1988[1]
Age group *(percentages)*				
18–24 years	*4.5*	*5.6*	*9.0*	*12.4*
25–49 years	*2.2*	*2.6*	*4.6*	*6.3*
18–49 years	*2.7*	*3.3*	*5.5*	*7.7*
Women in sample				
(= 100%) (numbers)				
18–24 years	1,353	1,517	1,194	1,215
25–49 years	4,651	5,007	4,320	4,250
18–49 years	6,004	6,524	5,514	5,465

1 1988-89 data. The General Household Survey changed from calendar years to financial years in 1988.

Source: General Household Survey

From: Social Trends, 1991, Table 2.16

2.12 Deaths by age and sex
United Kingdom   Rates and thousands

	Death rates per 1,000 in each age group							Total deaths (thousands)
	Under 1[1]	1-14	15-39	females 40-59 males 40-64	females 60-79 males 65-79	80 and over	All ages	
1961								
Males	24.8	0.6	1.3	11.8	66.1	190.7	12.6	322.0
Females	19.3	0.4	0.8	4.9	32.2	136.7	11.4	309.8
1971								
Males	20.2	0.5	1.1	11.4	59.9	174.0	12.1	328.5
Females	15.5	0.4	0.6	4.8	27.5	132.9	11.0	316.5
1976								
Males	16.4	0.4	1.1	11.1	60.4	183.4	12.5	341.9
Females	12.4	0.3	0.6	4.7	28.0	140.8	11.7	338.9
1981								
Males	12.7	0.4	1.0	10.1	56.1	167.5	12.0	329.1
Females	9.6	0.3	0.5	4.4	26.4	126.2	11.4	328.8
1986								
Males	10.9	0.3	0.9	9.1	54.0	158.2	11.8	327.2
Females	8.1	0.2	0.5	3.7	25.7	120.5	11.5	333.6
1988								
Males	10.2	0.3	1.0	8.3	51.1	147.1	11.5	319.1
Females	7.7	0.2	0.5	3.5	24.9	114.9	11.3	330.0
1989								
Males	9.5	0.3	1.0	7.8	50.5	149.6	11.5	320.2
Females	7.2	0.2	0.5	3.4	25.1	117.0	11.5	337.5

1 Rate per 1,000 live births.

Source: Office of Population Censuses and Surveys;
General Register Office (Scotland);
General Register Office (Northern Ireland)

From: Social Trends, 1991, Table 1.13

EMPLOYMENT 3

Definitions and sources

Definitional notes are given as footnotes to the tables which have been taken from the *Monthly Digest of Statistics*. The Monthly Digest tables refer to the United Kingdom unless otherwise stated. The charts have been taken from *Social Trends* and *Economic Trends*. More detailed statistics and commentary are published in the Employment Department's *Employment Gazette*.

For other sources see:

Guide to Official Statistics, 1990 edition (200 pages approximately fully indexed) HMSO.

Employment Gazette HMSO.

3.1 Distribution of the workforce

Thousands

			Not seasonally adjusted					Seasonally adjusted	
			Employees in employment[2]			Self-employed persons (with or without employees)[3]	HM Forces[4]		
	Workforce[1]	Workforce in employment[1]	Males	Females	Total			Workforce[1]	Employees in employment[2]
At June									
	DYDB	DYDA	BCAE	BCAF	BCAD	BCAG	BCAH	DYDD	BCAJ
1982	26 676	23 907	12 205	9 209	21 414	2 169	324	26 610	21 395
1983	26 608	23 624	11 940	9 127	21 067	2 219	322	26 633	21 054
1984	27 265	24 235	11 888	9 350	21 238	2 496	326	27 309	21 229
1985	27 718	24 539	11 903	9 521	21 423	2 614	326	27 743	21 414
1986	27 797	24 568	11 744	9 644	21 387	2 633	322	27 877	21 379
1988 Q1	28 223	25 631	11 893	10 115	22 008	2 963	317	28 293	22 104
Q2	28 255	25 914	11 971	10 287	22 258	2 998	316	28 347	22 266
Q3	28 503	26 192	12 046	10 401	22 446	3 062	315	28 387	22 405
Q4	28 460	26 413	11 986	10 580	22 566	3 126	313	28 369	22 496
1989 Q1	28 457	26 496	11 948	10 599	22 547	3 190	312	28 490	22 635
Q2	28 427	26 684	11 992	10 668	22 661	3 253	308	28 486	22 670
Q3	28 505	26 802	12 074	10 689	22 762	3 264	308	28 454	22 728
Q4	28 556	26 917	12 080	10 807	22 887	3 274	306	28 482	22 814
1990 Q1	28 387	26 742	12 015	10 701	22 716	3 284	306	28 436	22 802
Q2	28 436	26 881	12 050	10 806	22 855	3 298	303	28 509	22 864
Q3	28 512	26 838	12 068	10 755	22 823	3 298	303	28 481	22 793
Q4	28 591	26 740	11 919	10 796	22 715	3 298	300	28 512	22 641

1 The workforce consists of the workforce in employment and the unemployed (claimants); the workforce in employment comprises employees in employment, the self-employed, HM Forces and participants in work-related government training programmes. For more details see the August 1988 edition of *Employment Gazette*.

2 Estimates of employees in employment for periods after September 1989 and subsequent months include an allowance based on the Labour Force Survey to compensate for persistent undercounting in the regular sample enquiries *(Employment Gazette* April 1991, page 175).

3 Estimates of the self-employed up to mid-1990 are based on the 1981 census of population and the results of Labour Force Surveys carried out between 1981 and 1990. The figures for June 1990 are carried forward for later dates pending the results of the 1991 Labour Force Survey. A detailed description of the derivation of the estimates is given in the article in the April 1991 issue of *Employment Gazette*.

4 HM Forces figures, provided by the Ministry of Defence, represent the total number of UK service personnel, male and female, in HM Regular Forces, wherever serving and including those on release leave. The numbers are not subject to seasonal adjustment.

Sources: Department of Employment;
Department of Economic Development (Northern Ireland)

From: Monthly Digest of Statistics, June 1991, Table 3.1.

3.2 Change in the labour force: by age 1989-2001
Great Britain

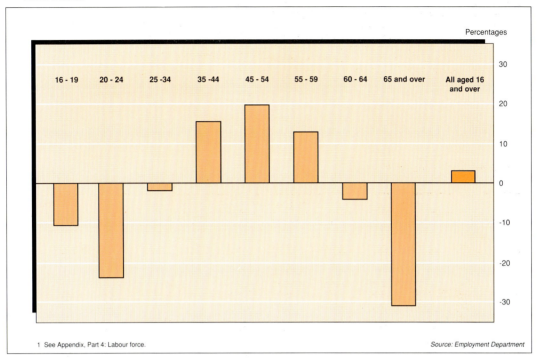

1 See Appendix, Part 4: Labour force.

Source: Employment Department

From: Social Trends 1991 Chart 4.5.

3.3 Population of working age: by sex and economic status
Great Britain

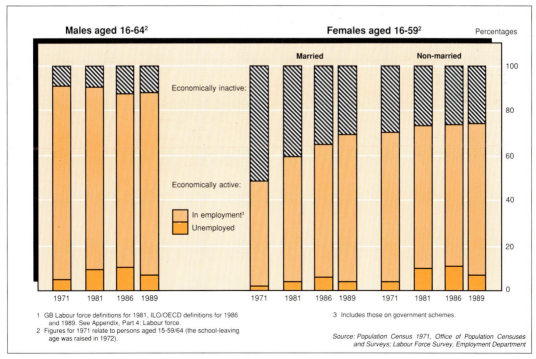

1 GB Labour force definitions for 1981, ILO/OECD definitions for 1986 and 1989. See Appendix, Part 4: Labour force.
2 Figures for 1971 relate to persons aged 15-59/64 (the school-leaving age was raised in 1972).

3 Includes those on government schemes.

Source: Population Census 1971, Office of Population Censuses and Surveys; Labour Force Survey, Employment Department

From: Social Trends 1991, Chart 4.2.

3.4 Regional unemployment rates

Percentage rates

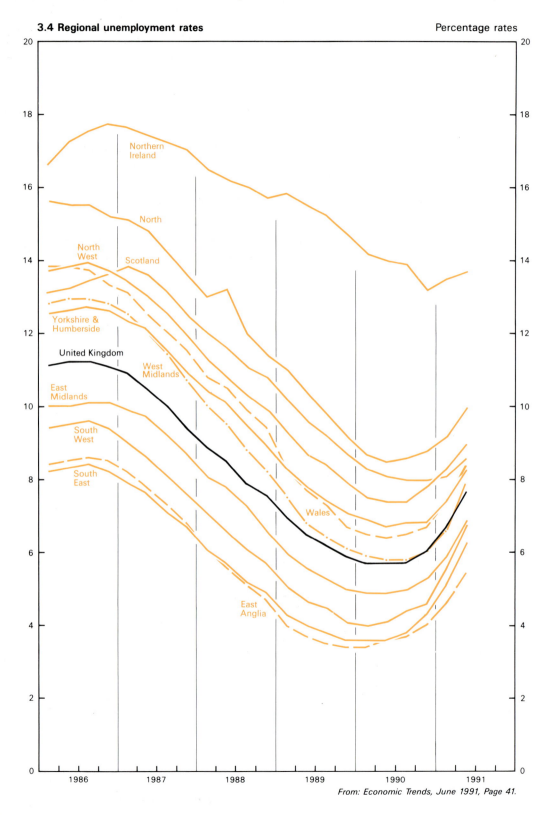

Northern Ireland

North

North West

Scotland

Yorkshire & Humberside

United Kingdom

West Midlands

East Midlands

South West

South East

Wales

East Anglia

From: Economic Trends, June 1991, Page 41.

3.5 Unemployment

Thousands

	United Kingdom						Great Britain	
	Not seasonally adjusted[1,2,3]		Seasonally adjusted[5]				Seasonally adjusted[5]	
	Total	Percentage rate [4]	Males	Females	Total	Percentage rate [4]	Total	Percentage rate [4]
	BCJA	BCJB	DPAE	DPAF	BCJD	BCJE	DPAG	DPAJ
1985	3 271.3	11.8	2 106.5	921.5	3 027.9	10.9	2 915.2	10.8
1986	3 289.1	11.8	2 139.0	959.0	3 097.9	11.1	2 975.3	11.0
1987	2 953.4	10.6	1 955.3	851.2	2 806.5	10.0	2 684.4	10.0
1988	2 370.4	8.4	1 588.1	686.8	2 274.9	8.1	2 161.7	8.0
1989	1 798.7	6.3	1 277.4	507.0	1 784.4	6.3	1 678.8	6.2
1990	1 664.5	5.8	1 230.3	431.4	1 661.7	5.8	1 564.6	5.6
1988 Apr	2 536.0	9.1	1 663.6	726.8	2 390.4	8.5	2 275.7	8.3
May	2 426.9	8.8	1 631.2	712.1	2 343.3	8.3	2 228.8	8.1
Jun	2 340.8	8.3	1 602.5	696.3	2 298.8	8.1	2 184.8	7.9
Jul	2 326.7	8.2	1 563.5	677.6	2 241.1	7.9	2 127.9	7.7
Aug	2 291.2	8.1	1 538.2	662.5	2 200.7	7.8	2 088.3	7.6
Sep	2 311.0	8.2	1 519.8	651.6	2 171.4	7.7	2 059.8	7.5
Oct	2 118.9	7.5	1 496.7	636.3	2 133.0	7.6	2 022.4	7.4
Nov	2 066.9	7.3	1 462.1	621.4	2 083.5	7.4	1 972.8	7.2
Dec	2 046.5	7.2	1 421.4	600.3	2 021.7	7.2	1 912.5	7.0
1989 Jan	2 074.3	7.3	1 395.2	586.4	1 981.6	7.0	1 871.7	6.8
Feb	2 018.2	7.1	1 366.3	571.0	1 937.3	6.8	1 827.7	6.6
Mar	1 960.2	6.9	1 346.7	556.5	1 903.2	6.7	1 794.2	6.5
Apr	1 883.6	6.6	1 312.5	534.3	1 846.8	6.5	1 738.8	6.3
May	1 802.5	6.3	1 295.0	524.0	1 819.0	6.4	1 711.9	6.2
Jun	1 743.1	6.1	1 279.6	511.6	1 791.2	6.3	1 685.3	6.1
Jul	1 771.4	6.2	1 265.7	500.5	1 766.2	6.2	1 660.4	6.0
Aug	1 741.1	6.1	1 243.1	481.9	1 725.0	6.1	1 620.4	5.8
Sep	1 702.9	6.0	1 218.6	466.1	1 684.7	5.9	1 581.7	5.7
Oct	1 635.8	5.7	1 211.2	459.2	1 670.4	5.9	1 568.1	5.7
Nov	1 612.4	5.7	1 200.0	451.1	1 651.1	5.8	1 549.9	5.6
Dec	1 639.0	5.8	1 194.7	441.4	1 636.1	5.8	1 535.7	5.5
1990 Jan	1 687.0	5.9	1 181.7	434.1	1 615.8	5.7	1 516.6	5.5
Feb	1 675.7	5.9	1 182.4	431.6	1 614.0	5.7	1 515.3	5.5
Mar	1 646.6	5.8	1 177.9	428.7	1 606.6	5.6	1 508.1	5.4
Apr	1 626.3	5.7	1 177.2	429.8	1 607.0	5.6	1 509.0	5.4
May	1 578.5	5.5	1 184.0	426.9	1 610.9	5.7	1 513.2	5.5
Jun	1 555.6	5.5	1 193.5	424.9	1 618.4	5.7	1 521.5	5.5
Jul	1 623.6	5.7	1 210.4	421.7	1 632.1	5.7	1 535.2	5.5
Aug	1 657.8	5.8	1 230.2	425.1	1 655.3	5.8	1 559.5	5.6
Sep	1 673.9	5.9	1 246.6	423.9	1 670.5	5.9	1 575.0	5.7
Oct	1 670.6	5.9	1 273.8	431.0	1 704.8	6.0	1 609.4	5.8
Nov	1 728.1	6.1	1 320.1	443.0	1 763.1	6.2	1 666.8	6.0
Dec	1 850.4	6.5	1 385.8	456.5	1 842.3	6.5	1 745.4	6.3
1991 Jan	1 959.7	6.9	1 425.6	466.0	1 891.6	6.7	1 794.2	6.5
Feb	2 045.4	7.2	1 495.6	484.2	1 979.8	7.0	1 882.2	6.8
Mar	2 142.1	7.5	1 581.2	509.8	2 091.0	7.4	1 992.2	7.2
Apr	2 198.5	7.7	1 644.8	528.8	2 173.6	7.6	2 074.4	7.5
May	2 213.8	7.8	1 699.5	544.7	2 244.2	7.9	2 144.8	7.7

1 Unadjusted figures from September 1988 are affected by the new benefit regulations for those aged under 18, most of whom are no longer eligible for Income Support. This reduces the UK unadjusted total by about 90 000 on average with most of this effect having taken place over the two months to October 1988.

2 The unadjusted unemployment figures between September 1989 and March 1990 are affected by the change in the conditions of the Reduntant Mineworkers Payment Scheme. An estimated 15 500 men left the count as a result of this change.

3 The unadjusted figures for September 8 1988 include some temporary over-recording, estimated at about 55 000, because of the postal strike in Great Britain (Northern Ireland was unaffected). An allowance for this distortion has been made in the seasonally adjusted figures for September.

4 Percentage rates have been calculated by expressing the total numbers unemployed as percentages of the numbers of employees in employment, unemployed, self-employed, HM Forces and participants on government training programmes at the appropriate mid-year.

5 The seasonally adjusted relate only to claimants aged 18 or over, in order to maintain the consistent series, available back to 1971 (1974 for the regions), allowing for the effect of the change in benefit regulations for under 18 year olds from September 1988 (see page 660 of the December 1988 *Employment Gazette*). The seasonally adjusted series also takes account of past discontinuities to be consistent with current coverage (see page 422 of the October 1986 *Employment Gazette* for the list of previous discontinuities taken into account).

Sources: Department of Employment;
Department of Economic Development (Northern Ireland)

From: Monthly Digest of Statistics, June 1991, Table 3.10.

Definitions and sources

National accounts provide a comprehensive and detailed framework for describing and analysing the economy as a whole and showing how various economic activities are related. They provide the basic background for decision-taking and forecasting in both Government and business. The accounts are published annually in *'United Kingdom and National Accounts: The CSO Blue Book'* and quarterly in regular articles in *'Economic Trends'.* A definitive detailed description of the various statistical series which comprise the national accounts is given in *'United Kingdom National Accounts: Sources and Methods'.* In general, the United Kingdom national accounts follow the principles recommended internationally.

Gross domestic product (GDP) is a concept of the value of goods and services produced on the economic territory irrespective of to whom the benefits accrue (United Kingdom residents or non-residents). The national product indicates the activities of UK residents only on both the UK economic territory and abroad.

The expenditure approach to gross domestic product, GDP(E) differentiates between consumption expenditure (goods and services consumed within a short time of purchase) and investment expenditure which adds to the domestic stock of physical assets (capital formation) or to those claims on non-residents which arise from the difference between exports and imports of goods and services. Estimates are compiled in both current and constant prices. The deflator implied by these current and constant price estimates at factor cost conceptually measures the price of domestic value added and is known as the 'index of total home costs'.

The income approach to gross domestic product GDP (I), identifies the different types of factor incomes derived from domestic production such as income from employment, from self-employment, profits and rent. It is compiled only in current price terms but a constant price equivalent is derived by deflating total incomes by the total home costs deflator.

The output approach to gross domestic product, GDP (O), provides estimates of the contribution of each industry. The production and construction industries account for about two-fifths of total output. Agriculture and the services industries, for example, distribution, transport, and financial services, make up the remainder. Estimates are available only at constant prices in index number form. The *' Monthly Digest of Statistics'* contains monthly detail for production industries.

Although estimates of GDP based on expenditure income and output should in principle give the same result, in practice there are often variations between them because of errors and omissions in the estimates. The output measure is usually the best indicator of quarter to quarter movements; for comparisons over periods of a year or more, the average of the three measures, GDP (A), is preferred.

Gross national disposable income at market prices is a concept of the United Kingdom's command over resources. It is based on the average measure of GDP at market prices, adjusted for the effect of changes in the terms of trade, for net property income from abroad and net current transfers abroad.

For other sources see:

Guide to Official Statistics: 1990 edition (200 pages approximately fully indexed) HMSO.

4.1 National and domestic product: average estimates

	1970	1980	1981	1982	1983	1984	1985	1986	1987	1988	1989	1990
THE MAIN AGGREGATES (1985=100)												
GDP(A) at current market prices ("money GDP")[1]	14.5	65.1	71.6	78.3	85.5	91.3	100.0	107.5	118.3	131.4	143.6	154.6
GDP(A) at 1985 factor cost	75.7	90.7	89.6	91.2	94.6	96.3	100.0	103.6	108.3	112.8	115.3	116.2
GNDI at 1985 market prices[2]	76.0	89.8	89.6	91.1	95.0	97.2	100.0	103.7	108.1	113.4	116.2	117.8
Index of total home costs[3]	19.1	72.3	79.6	85.2	90.0	94.8	100.0	102.7	107.9	114.9	123.4	134.1
AT CURRENT PRICES (£ million)												
At market prices												
Gross domestic product at market prices ("money GDP")[1]	51 770	231 772	254 851	278 887	304 314	325 091	356 083	382 942	421 198	467 863	511 413	550 597
Net property income from abroad	596	-182	1 251	1 460	2 831	4 357	2 646	5 096	4 078	5 047	4 088	4 029
Gross national product at market prices[1]	52 366	231 590	256 102	280 347	307 145	329 448	358 729	388 038	425 276	472 910	515 501	554 626
Net transfer income from abroad	-182	-1 984	-1 547	-1 741	-1 593	-1 730	-3 111	-2 157	-3 400	-3 518	-4 578	-4 935
Gross national disposable income[1]	52 184	229 606	254 555	278 606	305 552	327 718	355 618	385 881	421 876	469 392	510 923	549 691
At factor cost												
Gross domestic product at market prices	51 770	231 772	254 851	278 887	304 314	325 091	356 083	382 942	421 198	467 863	511 413	550 597
Adjustment to factor cost	-7 468	-30 755	-36 096	-40 656	-43 231	-45 039	-49 367	-56 760	-62 901	-70 571	-75 233	-72 850
Gross domestic product at factor cost	44 302	201 017	218 755	238 231	261 083	280 052	306 716	326 182	358 297	397 292	436 180	477 747
Net property income from abroad	596	-182	1 251	1 460	2 831	4 357	2 646	5 096	4 078	5 047	4 088	4 029
Gross national product at factor cost	44 898	200 835	220 006	239 691	263 914	284 409	309 362	331 278	362 375	402 339	440 268	481 776
less Capital consumption	-4 618	-27 952	-31 641	-33 653	-36 150	-38 758	-41 883	-45 084	-48 149	-52 828	-56 337	-61 159
Net national product at factor cost ("National income")	40 280	172 883	188 365	206 038	227 764	245 651	267 479	286 194	314 226	349 511	383 931	420 617
AT 1985 PRICES (£ million)												
At market prices												
Gross domestic product at market prices	266 997	323 419	319 193	324 622	336 503	343 780	356 083	370 030	387 718	404 230	413 467	416 888
Net property income from abroad	3 042	-255	1 627	1 774	3 203	4 532	2 646	5 328	4 153	5 171	3 936	3 798
Gross national product at market prices	270 586	323 178	320 826	326 402	339 706	348 312	358 729	375 358	391 871	409 401	417 403	420 686
Net transfer income from abroad	-929	-2 782	-2 012	-2 116	-1 802	-1 800	-3 111	-2 255	-3 462	-3 605	-4 407	-4 652
Terms of trade effect	595	-882	-353	-432	-22	-980	–	-4 266	-4 104	-2 656	288	2 779
Real national disposable income	270 429	319 519	318 462	323 855	337 882	345 532	355 618	368 837	384 305	403 140	413 284	418 813
At factor cost												
Gross domestic product at market prices	266 997	323 419	319 193	324 622	336 503	343 780	356 083	370 030	387 718	404 230	413 467	416 888
Adjustment to factor cost	-34 588	-45 305	-44 246	-44 895	-46 355	-48 347	-49 367	-52 312	-55 539	-58 312	-59 974	-60 556
Gross domestic product at factor cost	232 301	278 160	274 964	279 738	290 148	295 433	306 716	317 718	332 179	345 918	353 493	356 332
Net property income from abroad	3 042	-255	1 627	1 774	3 203	4 532	2 646	5 328	4 153	5 171	3 936	3 798
Gross national product at factor cost	235 712	277 946	276 608	281 528	293 351	299 965	309 362	323 046	336 332	351 089	357 429	360 130
less Capital consumption	-24 904	-36 417	-37 625	-38 744	-39 872	-40 916	-41 883	-42 552	-43 290	-44 368	-44 498	-45 518
Net national product at factor cost ("National income")	211 104	241 628	238 981	242 749	253 479	259 049	267 479	280 494	293 042	306 721	312 931	314 612
INDEX NUMBERS (1985=100)												
Value indices:												
Gross domestic product at market prices ("money GDP")[1]	14.5	65.1	71.6	78.3	85.5	91.3	100.0	107.5	118.3	131.4	143.6	154.6
Gross domestic product at factor cost	14.4	65.5	71.3	77.7	85.1	91.3	100.0	106.3	116.8	129.5	142.2	155.8
Volume indices:												
Gross domestic product at market prices	75.0	90.8	89.6	91.2	94.5	96.5	100.0	103.9	108.9	113.5	116.1	117.1
Gross national product at market prices	75.4	90.1	89.4	91.0	94.7	97.1	100.0	104.6	109.2	114.1	116.4	117.3
Gross national disposable income	76.0	89.8	89.6	91.1	95.0	97.2	100.0	103.7	108.1	113.4	116.2	117.8
Gross domestic product at factor cost	75.7	90.7	89.6	91.2	94.6	96.3	100.0	103.6	108.3	112.8	115.3	116.2
Gross national product at factor cost	76.2	89.8	89.4	91.0	94.8	97.0	100.0	104.4	108.7	113.5	115.5	116.4
Net national product at factor cost ("National income")	78.9	90.3	89.3	90.8	94.8	96.8	100.0	104.9	109.6	114.7	117.0	117.6
Price indices:[4]												
Gross domestic product at market prices[1]	19.4	71.7	79.8	85.9	90.4	94.6	100.0	103.5	108.6	115.7	123.7	132.1
Gross domestic product at factor cost ("Index of total home costs")	19.1	72.3	79.6	85.2	90.0	94.8	100.0	102.7	107.9	114.9	123.4	134.1

1 This series is affected by the abolition of domestic rates and the introduction of the community charge.
2 "Real national disposable income".
3 Expenditure-based deflators at factor cost.
4 Expenditure-based deflators.

From: United Kingdom National Accounts, 1991, Table 1.1

NATIONAL ACCOUNTS (GDP)

4.2 National product: by category of expenditure

£ million

	1970	1980	1981	1982	1983	1984	1985	1986	1987	1988	1989	1990
AT CURRENT MARKET PRICES:												
Consumers' expenditure[1]	32 114	139 608	155 412	170 650	187 028	199 425	217 618	241 275	264 880	298 796	326 489	349 421
General government final consumption	9 038	48 940	55 374	60 363	65 787	69 760	73 805	79 381	85 349	91 729	99 029	109 495
of which: Central Government	5 481	29 989	33 879	37 000	40 654	43 142	45 879	48 801	52 040	55 610	60 527	66 858
Local authorities	3 557	18 951	21 495	23 363	25 133	26 618	27 926	30 580	33 309	36 119	38 502	42 637
Gross domestic fixed capital formation	9 736	41 561	41 304	44 824	48 615	54 967	60 353	64 514	74 077	88 958	101 842	105 195
Value of physical increase in stocks and work in progress	382	-2 572	-2 768	-1 188	1 465	1 296	821	716	1 388	4 800	3 155	-718
Total domestic expenditure[1]	51 270	227 537	249 322	274 649	302 895	325 448	352 597	385 886	425 694	484 283	530 515	563 393
Exports of goods and services	11 510	62 616	67 432	72 694	80 056	91 852	102 208	98 319	107 031	107 834	122 791	134 108
of which: Goods	8 128	47 149	50 668	55 331	60 700	70 265	77 991	72 627	79 153	80 346	92 389	102 038
Services	3 382	15 467	16 764	17 363	19 356	21 587	24 217	25 692	27 878	27 488	30 402	32 070
Total final expenditure[1]	62 780	290 153	316 754	347 343	382 951	417 300	454 805	484 205	532 725	592 117	653 306	697 501
less Imports of goods and services[2]	-11 103	-57 606	-60 388	-67 762	-77 529	-92 669	-98 866	-101 070	-111 868	-124 884	-142 704	-147 582
of which: Goods	-8 142	-45 792	-47 416	-53 421	-62 237	-75 601	-81 336	-82 186	-90 735	-101 970	-116 987	-120 713
Services	-2 961	-11 814	-12 972	-14 341	-15 292	-17 068	-17 530	-18 884	-21 133	-22 914	-25 717	-26 869
Gross domestic product (expenditure-based)[1,3]	51 677	232 547	256 366	279 581	305 422	324 631	355 939	383 135	420 857	467 233	510 602	549 919
Statistical discrepancy (expenditure adjustment)[4]	93	-775	-1 515	-694	-1 108	460	144	-193	341	630	811	678
Gross domestic product (average estimate)[1,3]	51 770	231 772	254 851	278 887	304 314	325 091	356 083	382 942	421 198	467 863	511 413	550 597
Net property income from abroad	596	-182	1 251	1 460	2 831	4 357	2 646	5 096	4 078	5 047	4 088	4 029
Gross national product (average estimate)[1,3,5]	52 366	231 590	256 102	280 347	307 145	329 448	358 729	388 038	425 276	472 910	515 501	554 626
FACTOR COST ADJUSTMENT:												
Taxes on expenditure	8 352	36 474	42 465	46 467	49 500	52 576	56 592	62 947	69 074	76 511	80 925	79 067
Subsidies	884	5 719	6 369	5 811	6 269	7 537	7 225	6 187	6 173	5 940	5 692	6 217
Factor cost adjustment (taxes less subsidies)	7 468	30 755	36 096	40 656	43 231	45 039	49 367	56 760	62 901	70 571	75 233	72 850
AT CURRENT FACTOR COST:												
Consumers' expenditure	26 473	118 006	129 667	141 234	155 544	166 209	180 819	199 067	218 423	246 619	270 303	296 671
General government final consumption	8 401	45 348	51 197	55 734	61 455	65 540	69 559	74 569	80 169	86 163	93 068	102 878
Gross domestic capital formation	9 405	36 334	35 686	40 449	46 312	52 130	56 562	59 972	69 264	86 368	97 605	97 204
Total domestic expenditure	44 279	199 688	216 550	237 417	263 311	283 879	306 940	333 608	367 856	419 150	460 976	496 753
Exports of goods and services	11 033	59 710	64 108	69 270	76 409	88 382	98 498	93 837	101 968	102 396	117 097	127 898
Total final expenditure	55 312	259 398	280 658	306 687	339 720	372 261	405 438	427 445	469 824	521 546	578 073	624 651
less Imports of goods and services	-11 103	-57 606	-60 388	-67 762	-77 529	-92 669	-98 866	-101 070	-111 868	-124 884	-142 704	-147 582
Gross domestic product (expenditure-based)	44 209	201 792	220 270	238 925	262 191	279 592	306 572	326 375	357 956	396 662	435 369	477 069
Statistical discrepancy (expenditure adjustment)[4]	93	-775	-1 515	-694	-1 108	460	144	-193	341	630	811	678
Gross domestic product (average estimate)	44 302	201 017	218 755	238 231	261 083	280 052	306 716	326 182	358 297	397 292	436 180	477 747
Net property income from abroad	596	-182	1 251	1 460	2 831	4 357	2 646	5 096	4 078	5 047	4 088	4 029
Gross national product (average estimate)[5]	44 898	200 835	220 006	239 691	263 914	284 409	309 362	331 278	362 375	402 339	440 268	481 776
less Capital consumption	-4 618	-27 952	-31 641	-33 653	-36 150	-38 758	-41 883	-45 084	-48 149	-52 828	-56 337	-61 159
Net national product at factor cost (average estimate): "National income"[5]	40 280	172 883	188 365	206 038	227 764	245 651	267 479	286 194	314 226	349 511	383 931	420 617
VALUE INDICES, AT CURRENT PRICES (1985=100)												
Gross domestic product (expenditure-based)												
At market prices[1]	14.5	65.3	72.0	78.5	85.8	91.2	100.0	107.6	118.2	131.3	143.5	154.5
At factor cost	14.4	65.8	71.8	77.9	85.5	91.2	100.0	106.5	116.8	129.4	142.0	155.6
Gross domestic product (average estimate)												
At market prices ("money GDP")[1]	14.5	65.1	71.6	78.3	85.5	91.3	100.0	107.5	118.3	131.4	143.6	154.6
At factor cost	14.4	65.5	71.3	77.7	85.1	91.3	100.0	106.3	116.8	129.5	142.2	155.8

1 This series is affected by the abolition of domestic rates and the introduction of the community charge.
2 Excluding taxes on expenditure levied on imports.
3 Including taxes on expenditure levied on imports.
4 The Statistical discrepancy (expenditure adjustment) is part of the Residual error attributed to the expenditure-based estimate of GDP.

5 Blue Books before the 1988 edition showed expenditure-based estimates of Gross national product and of National income. These expenditure-based estimates can be derived as follows:

	1980	1981	1982	1983	1984	1985	1986	1987	1988	1989	1990
At current market prices											
Gross domestic product (expenditure-based)	232 547	256 366	279 581	305 422	324 631	355 939	383 135	420 857	467 233	510 602	549 919
plus Net property income from abroad	-182	1 251	1 460	2 831	4 357	2 646	5 096	4 078	5 047	4 088	4 029
Gross national product (expenditure-based)	232 365	257 617	281 041	308 253	328 988	358 585	388 231	424 935	472 280	514 690	553 948
At current factor cost											
Gross domestic product (expenditure-based)	201 792	220 270	238 925	262 191	279 592	306 572	326 375	357 956	396 662	435 369	477 069
plus Net property income from abroad	-182	1 251	1 460	2 831	4 357	2 646	5 096	4 078	5 047	4 088	4 029
Gross national product (expenditure-based)	201 610	221 521	240 385	265 022	283 949	309 218	331 471	362 034	401 709	439 457	481 098
less Capital consumption	-27 952	-31 641	-33 653	-36 150	-38 758	-41 883	-45 084	-48 149	-52 828	-56 337	-61 159
National income (expenditure-based)	173 658	189 880	206 732	228 872	245 191	267 335	286 387	313 885	348 881	383 120	419 939

From: United Kingdom National Accounts, 1991, Table 1.2

Key Data 91, © Crown copyright 1991

NATIONAL ACCOUNTS (GDP)

4.3 Gross domestic product at current factor costs: by category of income £ million

	1980	1981	1982	1983	1984	1985	1986	1987	1988	1989	1990
FACTOR INCOMES											
Income from employment	137 783	149 737	158 838	169 847	180 883	195 708	211 729	229 532	255 357	283 585	316 408
Income from self-employment[1]	18 141	19 980	22 140	24 750	27 804	30 116	34 769	39 383	44 835	51 605	57 661
Gross trading profits of companies[1,2]	27 861	27 341	31 176	39 528	44 656	51 767	47 049	59 315	63 950	66 203	62 916
Gross trading surplus of public corporations[1]	6 309	7 974	9 502	10 004	8 381	7 120	8 059	6 802	7 354	6 418	4 265
Gross trading surplus of general government enterprises[1]	180	236	216	50	-117	265	155	-75	-32	199	17
Rent[3]	14 243	16 366	17 700	18 857	19 804	21 792	23 826	25 772	29 292	32 091	38 433
Imputed charge for consumption of non-trading capital	2 116	2 351	2 426	2 498	2 619	2 830	3 068	3 307	3 634	4 005	4 278
Total domestic income[1]	206 633	223 985	241 998	265 534	284 030	309 598	328 655	364 036	404 390	444 106	483 978
less Stock appreciation	-6 391	-5 974	-4 276	-4 204	-4 513	-2 738	-1 790	-4 725	-6 212	-7 292	-6 391
Gross domestic product (income-based)	200 242	218 011	237 722	261 330	279 517	306 860	326 865	359 311	398 178	436 814	477 587
Statistical discrepancy (income adjustment)[4]	775	744	509	-247	535	-144	-683	-1 014	-886	-634	160
Gross domestic product (average estimate) at factor cost	201 017	218 755	238 231	261 083	280 052	306 716	326 182	358 297	397 292	436 180	477 747
FACTOR INCOMES AFTER PROVIDING FOR STOCK APPRECIATION											
Income from employment	137 783	149 737	158 838	169 847	180 883	195 708	211 729	229 532	255 357	283 585	316 408
Income from self-employment[3]	17 422	19 354	21 778	24 200	27 479	29 641	34 603	38 893	44 091	50 784	56 996
Gross trading profits of companies[2,3]	22 467	22 277	27 665	35 909	40 533	49 612	45 539	55 179	58 724	59 816	57 304
Gross trading surplus of public corporations[3]	6 031	7 690	9 099	9 969	8 316	7 012	7 945	6 703	7 112	6 334	4 151
Gross trading surplus of general government enterprises[3]	180	236	216	50	-117	265	155	-75	-32	199	17
Rent[3]	14 243	16 366	17 700	18 857	19 804	21 792	23 826	25 772	29 292	32 091	38 433
Imputed charge for consumption of non-trading capital	2 116	2 351	2 426	2 498	2 619	2 830	3 068	3 307	3 634	4 005	4 278
Gross domestic product (income-based)	200 242	218 011	237 722	261 330	279 517	306 860	326 865	359 311	398 178	436 814	477 587
Statistical discrepancy (income adjustment)[4]	775	744	509	-247	535	-144	-683	-1 014	-886	-634	160
Gross domestic product (average estimate) at factor cost	201 017	218 755	238 231	261 083	280 052	306 716	326 182	358 297	397 292	436 180	477 747
VALUE INDICES (1985 = 100)											
At factor cost											
Gross domestic product (income-based)	65.3	71.0	77.5	85.2	91.1	100.0	106.5	117.1	129.8	142.3	155.6
Gross domestic product (average estimate)	65.5	71.3	77.7	85.1	91.3	100.0	106.3	116.8	129.5	142.2	155.8

1 Before providing for depreciation and stock appreciation.
2 Including financial institutions.
3 Before providing for depreciation.
4 The statistical discrepancy (income adjustment) is part of the Residual error attributed to the income-based estimate of GDP.

From: United Kingdom National Accounts, 1991, Table 1.3

NATIONAL ACCOUNTS (GDP)

4.4 Gross national product by category of expenditure at 1985 prices [1]

£ million at 1985 prices

	1980	1981	1982	1983	1984	1985	1986	1987	1988	1989	1990
AT 1985 MARKET PRICES											
Consumers' expenditure	195 825	196 011	197 980	206 932	210 254	217 618	231 172	243 279	261 330	270 575	273 304
General government final consumption[2]	70 872	71 086	71 672	73 089	73 792	73 805	75 106	76 034	76 486	77 182	79 371
of which: Central Government	43 690	44 108	44 421	45 281	45 741	45 879	46 684	46 753	46 942	47 363	48 415
Local authorities	27 160	26 976	27 228	27 808	28 051	27 926	28 422	29 281	29 544	29 819	30 956
Gross domestic fixed capital formation	53 416	48 298	50 915	53 476	58 034	60 353	61 813	67 753	76 648	81 845	79 893
Value of physical increase in stocks and work in progress	-3 371	-3 200	-1 281	1 357	1 084	821	737	1 158	4 031	2 668	-705
Total domestic expenditure	316 602	311 634	319 028	334 854	343 164	352 597	368 828	388 224	418 495	432 270	431 863
Exports of goods and services	88 726	88 064	88 798	90 589	96 525	102 208	107 052	113 094	113 150	117 929	123 642
of which: Goods	65 575	64 956	66 789	68 344	73 887	77 991	81 289	85 516	87 027	91 163	97 207
Services	23 247	23 213	22 017	22 245	22 638	24 217	25 763	27 578	26 123	26 766	26 435
Total final expenditure	405 285	399 644	407 791	425 443	439 689	454 805	475 880	501 318	531 645	550 199	555 505
less Imports of goods and services[3]	-80 781	-78 522	-82 348	-87 709	-96 394	-98 866	-105 662	-113 916	-127 964	-137 389	-139 123
of which: Goods	-64 078	-61 531	-64 983	-70 789	-78 839	-81 336	-87 326	-93 782	-106 027	-114 387	-115 751
Services	-16 707	-16 996	-17 369	-16 920	-17 555	-17 530	-18 336	-20 134	-21 937	-23 002	-23 372
Gross domestic product (expenditure-based)[4]	324 490	321 103	325 434	337 734	343 295	355 939	370 218	387 402	403 681	412 810	416 382
Statistical discrepancy[5] (expenditure adjustment)	-1 072	-1 904	-815	-1 231	485	144	-188	316	549	657	506
Gross domestic product (average estimate)[4]	323 419	319 193	324 622	336 503	343 780	356 083	370 030	387 718	404 230	413 467	416 888
Net property income from abroad	-255	1 627	1 774	3 203	4 532	2 646	5 328	4 153	5 171	3 936	3 798
Gross national product (average estimate)[4,7]	323 178	320 826	326 402	339 706	348 312	358 729	375 358	391 871	409 401	417 403	420 686
AT 1985 FACTOR COST											
Gross domestic product at market prices (expenditure-based)[4]	324 490	321 103	325 434	337 734	343 295	355 939	370 218	387 402	403 681	412 810	416 382
Statistical discrepancy (expenditure adjustment)[5]	-1 072	-1 904	-815	-1 231	485	144	-188	316	549	657	506
Gross domestic product at market prices (average estimate)[4]	323 419	319 193	324 622	336 503	343 780	356 083	370 030	387 718	404 230	413 467	416 888
less Factor cost adjustment[6]	-45 305	-44 246	-44 895	-46 355	-48 347	-49 367	-52 312	-55 539	-58 312	-59 974	-60 556
Gross domestic product at factor cost (average estimate)	278 160	274 964	279 738	290 148	295 433	306 716	317 718	332 179	345 918	353 493	356 332
Net property income from abroad	-255	1 627	1 774	3 203	4 532	2 646	5 328	4 153	5 171	3 936	3 798
Gross national product at factor cost (average estimate)[7]	277 946	276 608	281 528	293 351	299 965	309 362	323 046	336 332	351 089	357 429	360 130
less Capital consumption	-36 417	-37 625	-38 744	-39 872	-40 916	-41 883	-42 552	-43 290	-44 368	-44 498	-45 518
Net national product at factor cost (average estimate): "National income"[7]	241 628	238 981	242 749	253 479	259 049	267 479	280 494	293 042	306 721	312 931	314 612

1 For the years before 1983, totals differ from the sum of their components.
2 An analysis of general government consumption by function is given in the CSO *Blue Book*.
3 Excluding taxes on expenditure levied on imports.
4 Including taxes on expenditure levied on imports.

5 The difference between the average estimate of Gross domestic product and the expenditure-based measure at 1985 prices.
6 This represents taxes on expenditure *less* subsidies valued at constant rates.
7 Blue Books before the 1988 edition showed expenditure-based estimates of Gross national product and Net national product. These expenditure-based measures can be derived as follows:

	1980	1981	1982	1983	1984	1985	1986	1987	1988	1989	1990
At 1985 market prices											
Gross domestic product at market prices (expenditure-based)	324 490	321 103	325 434	337 734	343 295	355 939	370 218	387 402	403 681	412 810	416 382
plus Net property income from abroad	-255	1 627	1 774	3 203	4 532	2 646	5 328	4 153	5 171	3 936	3 798
Gross national product at market prices (expenditure-based)	324 248	322 736	327 214	340 937	347 827	358 585	375 546	391 555	408 852	416 746	420 180
At 1985 factor cost											
Gross domestic product at market prices (expenditure-based)	324 490	321 103	325 434	337 734	343 295	355 939	370 218	387 402	403 681	412 810	416 382
less Factor cost adjustment	-45 305	-44 246	-44 895	-46 355	-48 347	-49 367	-52 312	-55 539	-58 312	-59 974	-60 556
Gross domestic product at factor cost (expenditure-based)	279 232	276 868	280 553	291 379	294 948	306 572	317 906	331 863	345 369	352 836	355 826
plus Net property income from abroad	-255	1 627	1 774	3 203	4 532	2 646	5 328	4 153	5 171	3 936	3 798
Gross national product at factor cost (expenditure-based)	279 017	278 512	282 343	294 582	299 480	309 218	323 234	336 016	350 540	356 772	359 624
less Capital consumption	-36 417	-37 625	-38 744	-39 872	-40 916	-41 883	-42 552	-43 290	-44 368	-44 498	-45 518
Net national product at factor cost (expenditure-based)	242 698	240 892	243 561	254 710	258 564	267 335	280 682	292 726	306 172	312 274	314 106

From: United Kingdom National Accounts, 1991, Table 1.6

4.5 Gross domestic product at constant factor cost: by industry of output

1985 = 100

	Weight per 1000[1]											
	1985	1980	1981	1982	1983	1984	1985	1986	1987	1988	1989	1990
AGRICULTURE, FORESTRY AND FISHING	19	83.0	85.2	92.3	87.3	104.7	100.0	97.1	97.9	97.4	101.2	104.4
PRODUCTION:												
Energy and water supply:												
Coal and coke	12	139.9	136.2	130.4	125.2	55.9	100.0	114.2	110.8	109.9	105.8	97.3
Extraction of mineral oil and natural gas	62	66.2	73.0	83.2	91.1	97.4	100.0	101.2	98.6	90.1	73.4	73.4
Mineral oil processing	4	101.6	94.5	94.2	96.7	99.6	100.0	100.9	102.1	109.4	112.0	111.1
Other energy and water supply	28	92.4	94.6	93.9	97.2	82.4	100.0	109.9	112.9	113.8	115.0	116.0
Total energy and water supply	106	82.6	86.5	91.6	96.8	88.8	100.0	105.0	103.9	99.3	89.6	88.8
Manufacturing:												
Metals	9	90.0	95.4	92.8	94.2	92.9	100.0	100.3	108.6	122.3	124.7	121.3
Other minerals and mineral products	12	103.1	91.8	93.7	96.8	100.4	100.0	101.3	106.8	117.3	120.1	113.4
Chemicals	24	83.8	83.7	84.3	90.6	96.4	100.0	101.7	109.0	114.4	119.5	118.2
Man-made fibres	1	128.8	109.4	87.5	100.6	104.9	100.0	103.6	109.9	107.8	114.8	117.3
Metal goods nes	13	101.4	93.5	94.1	96.5	104.4	100.0	99.4	103.4	111.5	113.5	110.9
Mechanical engineering	29	108.3	96.7	98.1	94.3	96.0	100.0	96.5	96.8	105.3	109.7	111.8
Electrical and instrument engineering	34	77.8	73.0	76.4	84.4	94.1	100.0	100.6	106.3	117.9	126.2	125.2
Motor vehicles and parts	13	113.3	94.0	90.8	95.2	93.5	100.0	96.9	103.9	119.1	125.3	121.0
Other transport equipment including aerospace	13	109.5	113.4	110.5	104.0	99.5	100.0	111.9	112.6	107.8	127.7	130.1
Food	23	95.1	94.3	97.6	98.7	99.6	100.0	101.3	103.2	105.1	105.1	105.2
Drink and tobacco	8	107.0	103.3	100.7	101.7	102.7	100.0	99.6	103.3	103.6	106.9	108.6
Textiles	7	101.9	93.6	91.3	93.4	96.2	100.0	100.2	104.6	101.8	96.9	92.1
Clothing, footwear and leather	9	94.4	88.4	87.6	91.9	95.9	100.0	101.0	103.0	102.1	99.5	98.6
Paper, printing and publishing	24	101.2	96.1	92.8	93.4	97.8	100.0	104.2	114.4	125.2	132.0	133.8
All other manufacturing (including timber, furniture, rubber and plastics)	19	98.2	89.4	88.2	93.8	99.1	100.0	105.0	115.7	128.5	132.6	132.4
Total manufacturing	238	96.8	91.0	91.2	93.8	97.4	100.0	101.3	106.6	114.1	118.9	118.3
Total production	344	92.6	89.6	91.4	94.7	94.8	100.0	102.4	105.7	109.5	109.9	109.2
CONSTRUCTION	59	89.9	82.9	89.4	95.1	99.6	100.0	104.5	112.7	122.9	130.4	131.8
SERVICE INDUSTRIES:												
Distribution, hotels and catering; repairs	134	87.3	86.0	87.6	91.5	96.2	100.0	104.6	111.4	117.9	121.8	122.6
Transport and communication:												
Transport	43	94.0	93.2	90.1	92.1	95.2	100.0	102.2	112.1	119.7	125.5	126.3
Communication	27	83.0	84.8	87.4	90.8	97.6	100.0	107.6	113.2	121.2	129.9	135.8
Total transport and communication	70	89.7	89.9	89.0	91.6	96.1	100.0	104.3	112.5	118.5	125.3	128.0
Banking, finance, insurance, business services and leasing	155	73	76	81	87	94	100	111	122	133	139	144
Ownership of dwellings	59	94	96	97	98	99	100	100	100	101	102	103
Public administration, national defence and compulsory social security	71	102	102	100	100	100	100	100	100	99	99	101
Education and health services	85	95	96	97	99	99	100	101	104	107	108	108
Other services[2]	59	87	86	86	89	95	100	105	113	118	121	123
Adjustment for financial services	55	73	76	81	88	94	100	114	127	137	143	145
Total services	578	88.9	89.2	90.5	93.4	97.0	100.0	103.9	109.4	114.0	117.4	119.4
Gross domestic product (output-based)	1 000	90.7	89.3	91.1	94.1	96.7	100.0	103.3	108.1	112.7	115.3	116.4

1 The weights are proportional to the distribution of net output in 1985 and are used to combine the indices from 1983 onwards. For the method of calculation in earlier years see paragraph 5.19 of *United Kingdom National Accounts: Sources and Methods*, Third edition.

2 Comprising classes 92, 94, 96-99 and 00 of the Standard Industrial Classification Revised 1980.

From: United Kingdom National Accounts, 1991, Table 2.4

NATIONAL ACCOUNTS (GDP)

4.6 Value, volume and price indices

1985 = 100

	1980	1981	1982	1983	1984	1985	1986	1987	1988	1989	1990
VALUE INDICES AT CURRENT PRICES											
Average estimates											
Gross domestic product at market prices											
("money GDP")[1]	65.1	71.6	78.3	85.5	91.3	100.0	107.5	118.3	131.4	143.6	154.6
Gross domestic product at factor cost	65.5	71.3	77.7	85.1	91.3	100.0	106.3	116.8	129.5	142.2	155.8
Expenditure-based measure											
Gross domestic product at market prices[1]	65.3	72.0	78.5	85.8	91.2	100.0	107.6	118.2	131.3	143.5	154.5
Gross domestic product at factor cost	65.8	71.8	77.9	85.5	91.2	100.0	106.5	116.8	129.4	142.0	155.6
Income-based measure											
Gross domestic product at factor cost	65.3	71.0	77.5	85.2	91.1	100.0	106.5	117.1	129.8	142.3	155.6
VOLUME INDICES AT 1985 PRICES											
Average estimates											
At market prices											
Gross domestic product	90.8	89.6	91.2	94.5	96.5	100.0	103.9	108.9	113.5	116.1	117.1
Gross national product	90.1	89.4	91.0	94.7	97.1	100.0	104.6	109.2	114.1	116.4	117.3
Gross national disposable income ("RNDI")	89.8	89.6	91.1	95.0	97.2	100.0	103.7	108.1	113.4	116.2	117.8
At factor cost											
Gross domestic product	90.7	89.6	91.2	94.6	96.3	100.0	103.6	108.3	112.8	115.3	116.2
Gross national product	89.8	89.4	91.0	94.8	97.0	100.0	104.4	108.7	113.5	115.5	116.4
Net national product ("National income")	90.3	89.3	90.8	94.8	96.8	100.0	104.9	109.6	114.7	117.0	117.6
Categories of expenditure:											
At market prices											
Consumers' expenditure	90.0	90.1	91.0	95.1	96.6	100.0	106.2	111.8	120.1	124.3	125.6
General government final consumption	96.0	96.3	97.1	99.0	100.0	100.0	101.8	103.0	103.6	104.6	107.5
of which: Central Government	95.2	96.1	96.8	98.7	99.7	100.0	101.8	101.9	102.3	103.2	105.5
Local authorities	97.3	96.6	97.5	99.6	100.4	100.0	101.8	104.9	105.8	106.8	110.9
Gross domestic fixed capital formation	88.5	80.0	84.4	88.6	96.2	100.0	102.4	112.3	127.0	135.6	132.4
Total domestic expenditure	89.8	88.4	90.5	95.0	97.3	100.0	104.6	110.1	118.7	122.6	122.5
Exports of goods and services	86.8	86.2	86.9	88.6	94.4	100.0	104.7	110.7	110.7	115.4	121.0
of which: Goods	84.1	83.3	85.6	87.6	94.7	100.0	104.2	109.7	111.6	116.9	124.7
Services	96.0	95.9	90.9	91.9	93.5	100.0	106.4	113.9	107.9	110.5	109.2
Total final expenditure	89.1	87.9	89.7	93.5	96.7	100.0	104.6	110.2	116.9	121.0	122.1
Imports of goods and services	81.7	79.4	83.3	88.7	97.5	100.0	106.9	115.2	129.4	139.0	140.7
of which: Goods	78.8	75.6	79.9	87.0	96.9	100.0	107.4	115.3	130.3	140.6	142.3
Services	95.3	97.0	99.1	96.5	100.1	100.0	104.6	114.9	125.2	131.2	133.4
Gross domestic product (expenditure-based)	91.2	90.2	91.4	94.9	96.4	100.0	104.0	108.8	113.4	116.0	117.0
at market prices											
Factor cost adjustment	91.8	89.6	90.9	93.9	97.9	100.0	106.0	112.5	118.1	121.5	122.7
Expenditure-based measure											
Gross domestic product at factor cost	91.1	90.3	91.5	95.0	96.2	100.0	103.7	108.2	112.7	115.1	116.1
Income-based measure[2]											
Gross domestic product at factor cost	90.3	89.3	91.0	94.6	96.1	100.0	103.8	108.6	113.0	115.4	116.1
Output-based measure											
Gross domestic product at factor cost	90.7	89.3	91.1	94.1	96.7	100.0	103.3	108.1	112.7	115.3	116.4
PRICE INDICES (IMPLIED DEFLATORS)[3]											
Categories of expenditure:											
Consumers' expenditure	71.3	79.3	86.2	90.4	94.8	100.0	104.4	108.9	114.3	120.7	127.9
General government final consumption	69.1	77.9	84.2	90.0	94.5	100.0	105.7	112.3	119.9	128.3	138.0
Gross domestic fixed capital formation	77.8	85.5	88.0	92.9	94.7	100.0	104.4	109.3	116.1	124.4	131.7
Total domestic expenditure	71.9	80.0	86.1	90.5	94.8	100.0	104.6	109.7	115.7	122.7	130.5
Exports of goods and services	70.6	76.6	81.9	88.4	95.2	100.0	91.8	94.6	95.3	104.1	108.5
Total final expenditure	71.6	79.3	85.2	90.0	94.9	100.0	101.7	106.3	111.4	118.7	125.6
Imports of goods and services	71.3	76.9	82.3	88.4	96.1	100.0	95.7	98.2	97.6	103.9	106.1
Gross domestic product (expenditure-based)	71.7	79.8	85.9	90.4	94.6	100.0	103.5	108.6	115.7	123.7	132.1
at market prices											
HOME COSTS PER UNIT OF OUTPUT[4]											
Total home costs[5]	72.3	79.6	85.2	90.0	94.8	100.0	102.7	107.9	114.9	123.4	134.1
Income from employment	78.0	85.7	89.2	91.7	96.2	100.0	104.3	108.0	115.5	125.6	139.3
Gross profits and other income	62.2	68.8	78.0	87.0	92.3	100.0	99.8	107.6	113.7	119.5	124.9

1 This series is affected by the abolition of domestic rates and the introduction of the community charge.
2 Income data deflated by the implied GDP deflator at factor cost, based on expenditure data.
3 Implied deflators are given by dividing the estimates for each component at current market prices by the corresponding estimate at constant market prices.

4 These index numbers show how employment and trading incomes relate to the index of total home costs as explained in paragraphs 4.42-4.45 of *United Kingdom National Accounts: Sources and Methods*, third edition.
5 Gross domestic product at factor cost (expenditure-based) deflator.

From: United Kingdom National Accounts, 1991, Table 1.7

U.K. FINANCE

Definitions and sources

There is no single, universally accepted, definition of money. Four of the more widely used series are shown here. They and others are shown together with details of their composition, in section 11 of *Financial Statistics*.

The public sector borrowing requirement (PSBR) indicates the extent to which the public sector borrows from other sectors of the economy and overseas to finance the balance of expenditure and receipts arriving from its various activities.

A much wider range of financial series is shown in *Financial Statistics* monthly and the *Bank of England Quarterly Bulletin*. Further definitons are given in the supplementary notes to the *Monthly Digest of Statistics* and in the *Financial Statistics Explanatory Handbook*.

For other sources see:

Guide to Official Statistics, 1990 edition (200 pages approximately fully indexed) HMSO.

Financial Statistics HMSO.

5.1 Monetary aggregates

£ million

	Amount outstanding					
	'Narrow' money		'Broad' money			
	M0-the wide monetary base		M2		M4	
	Not seasonally adjusted	Seasonally adjusted	Not seasonally adjusted	Seasonally adjusted	Not seasonally adjusted	Seasonally adjusted
	AVAD	AVAE	AUYC	AUYG	AUYM	AUYN
1984	14 615	13 781	132 972	132 007	199 399	199 641
1985	15 161	14 315	145 701	144 686	225 293	225 575
1986	16 098	15 027	167 119	166 036	261 073	261 071
1987	16 633	15 664	185 468	184 263	303 755	303 251
1988	18 040	16 871	214 968	213 358	357 023	356 088
1989	19 006	17 820	236 256	234 216	425 248	424 514
1990	19 484	18 296	255 144	252 710	476 737	475 982
1988 Q3	16 793	16 717	209 125	207 775	346 160	343 450
Q4	18 040	16 871	214 968	213 358	357 023	356 088
1989 Q1	16 815	16 969	218 884	218 974	372 144	371 252
Q2	17 089	17 173	224 212	223 069	390 879	388 820
			224 352	223 209	390 824	388 765
Q3	17 504	17 419	230 008	228 825	409 314	405 852
Q4	19 006	17 820	236 256	234 216	425 248	424 514
1990 Q1	17 600	18 050	239 070	238 814	441 736	440 515
Q2	18 194	18 265	245 512	244 448	459 321	456 718
Q3	18 319	18 206	250 024	248 847	469 904	465 605
Q4	19 484	18 296	255 144	252 710	476 737	475 982
1991 Q1	18 153	18 516	263 347	262 761	485 551	484 026
1990 Mar	17 600	18 050	239 070	238 814	441 736	440 515
Apr	18 264	18 309	241 923	242 498	443 743	445 844
May	18 241	18 244	241 952	243 041	449 108	451 296
Jun	18 194	18 265	245 512	244 448	459 321	456 718
Jul	18 365	18 195	246 816	246 347	459 924	459 217
Aug	18 573	18 267	246 577	247 290	463 382	463 147
Sep	18 319	18 206	250 024	248 847	469 904	465 605
Oct	18 217	18 256	251 312	250 778	469 456	469 552
Nov	18 224	18 253	250 848	250 959	475 049	474 541
Dec	19 484	18 296	255 144	252 710	476 737	475 982
1991 Jan	18 437	18 406	256 657	257 631	476 474	478 387
Feb	17 961	18 488	257 273	259 679	478 208	481 988
Mar	18 153	18 516	263 347	262 761	485 551	484 026
Apr	18 361	18 606	265 461	265 855	488 083	490 560

Source: Bank of England

From: Monthly Digest of Statistics, June 1991, Table 17.4

U.K. FINANCE

5.2 Public sector borrowing requirement [1]

£ million

	Total		Contributions by:			Financed by:			Other private sector	
						Banks and building societies/Overseas sector				
						External finance				
	Not seasonally adjusted	Seasonally adjusted [2]	Central government [3]	Local authorities	Public corporations	Borrowing in sterling from banks	Foreign currency borrowing from banks	Other external finance	Notes and coin	Other
	ABEN	ABFP	ABEA	-AAZK	-AAZL	AQXV	AQXW	ABGH	AQUP	AQGG
1987	-1 436	-1 671	4 059	4 664	831	-1 252	-365	-6 234	708	8 018
1988	-11 868	-11 960	-4 933	4 223	2 712	-685	-572	-734	1 040	-7 170
1989	-9 310	-9 434	-5 135	1 994	2 181	-3 590	-46	3 894	897	-9 800
1990	-2 062	-2 068	-4 605	-3 152	609	540	4	-4 644	-160	1 178
Financial years										
1988/89	-14 659	-14 659	-7 119	4 604	2 936	-4 003	8	793	407	-8 021
1989/90	-7 966	-7 966	-5 630	943	1 393	-632	-85	1 366	849	-9 532
1990/91	-469	-469	-2 483	-2 043	29	..	..	..	..	..
1989 Q3	-427	-1 494	349	676	100	1 002	371	747	101	-2 632
Q4	-3 196	-2 538	-3 234	-58	20	-1 322	-295	1 581	695	-3 142
1990 Q1	-4 246	-2 711	-4 698	-1 186	734	-1 611	-225	-1 994	-338	-1 130
Q2	5 847	4 720	4 252	-1 647	52	1 337	62	346	420	2 299
Q3	-475	-1 550	-559	105	-189	-1 038	8	384	-126	245
Q4	-3 188	-2 527	-3 600	-424	12	1 852	159	-3 380	-116	-241
1991 Q1	-2 653	-1 112	-2 576	-77	154	..	..	..	..	..

1 For further details see *Financial Statistics* Tables 2.3, 2.5 and 2.6.
2 Financial year constrained.
3 An increase in debt is shown positive.

Source: Central Statistical Office

From: Monthly Digest of Statistics, June 1991, Table 17.2

5.3 Selected interest rates, exchange rates and security prices

	Selected retail banks' base rate	Average discount rate on Treasury bills	Inter-bank 3 month rate	British government securities 20 years yield [1]	Sterling exchange rate index 1985=100	Exchange rate US spot	Ordinary share price index [2]
		AJNB		AJLX	AJHV	AJGA	AJMA
1990 May	15.00	14.46	15.03-15.09	11.49	88.0	1.6775	1 100.35
Jun	15.00	14.32	14.94-14.97	11.01	90.4	1.7440	1 170.76
Jul	15.00	14.33	14.94-15.00	11.03	93.5	1.8592	1 162.07
Aug	15.00	14.29	14.97-15.00	11.41	95.3	1.8907	1 075.21
Sep	15.00	14.23	14.94-14.97	11.32	93.8	1.8705	1 008.65
Oct	14.00	13.13	13.78-13.81	11.12	94.8	1.9445	1 009.41
Nov	14.00	12.69	13.50-13.56	10.94	94.2	1.9370	1 007.00
Dec	14.00	13.06	14.00-14.00	10.40	93.3	1.9295	1 040.25
1991 Jan	14.00	12.81	13.97-13.97	10.22	94.1	1.9590	1 010.79
Feb	13.00	12.06	12.75-12.88	9.89	94.3	1.9135	1 095.30
Mar	12.50	11.56	12.38-12.44	10.06	92.9	1.7385	1 190.90
Apr	12.00	11.17	11.78-11.81	9.99	92.3	1.7205	1 217.27
May	11.50	10.84	11.38-11.41	10.15	91.7	1.7030	1 203.75

1 Average of working days.
2 *Financial Times* Actuaries share indices 10 April 1962 = 100. All classes (750 shares) index.

Source: Bank of England

From: Monthly Digest of Statistics, June 1991, Table 17.5

5.4 Selected financial statistics [1]

£ million

| | Building societies | | | | Unit trusts | Total capital issues (net) | Net inflow into life assurance & super-annuation funds |
| | Deposits | | Advances | | | | |
National savings[2]	Not seasonally adjusted	Seasonally adjusted	Not seasonally adjusted	Seasonally adjusted			
Amount outstanding							
31 Dec							
ACUV	AHIX		AHIF		AGXB		
1990 37 320	161 516		182 454		46 342		
Net transactions							
ACVX	AHKB	AHHR	AAMN	AHHU	AGXE	AJAD	AALV
1987 2 523	14 409	..	15 461	..	6 330	15 386	21 083
1988 1 492	20 685	..	24 926	..	1 796	7 062	21 488
1989 -1 489	17 517	..	26 460	..	3 864	7 863	30 157
1990 972	18 052	..	26 256	..	393	2 829	31 255
1990 Q1 -220	4 069	3 746	6 828	7 498	415	1 293	7 192
Q2 287	4 249	4 737	6 805	6 556	144	498	7 184
Q3 467	4 531	4 213	6 626	6 109	-590	793	8 775
Q4 438	5 203	5 357	5 997	6 094	424	245	8 104
1991 Q1 299	6 135	5 634	..	5 729	1 274	130	7 535
1990 Jun 76	1 893	1 951	2 446	2 105	-64	289	..
Jul 261	2 188	1 330	2 334	2 189	203	443	..
Aug 115	643	1 344	2 242	1 936	-476	313	..
Sep 91	1 700	1 539	2 050	1 984	-317	37	..
Oct 195	1 448	1 846	2 163	2 066	189	10	..
Nov 128	-43	1 498	2 368	2 178	51	-175	..
Dec 115	3 798	2 013	1 466	1 850	184	410	..
1991 Jan 135	3 730	1 888	1 681	2 053	130	50	..
Feb 54	941	1 825	..	1 889	836	537	..
Mar 110	1 464	1 921	..	1 787	308	-457	..
Apr 253	2 367	2 336	..	1 865	217	1 090	..
May ..	..	..	..	..	..	1 443	..

	Banks[3]						Credit business: Total agreements		Consumer credit	
	UK private sector deposits			Lending to the private sector						
	Sterling			Sterling						
	Not seasonally adjusted	Seasonally adjusted	Other currencies	Not seasonally adjusted	Seasonally adjusted	Other currencies	Not seasonally adjusted	Seasonally adjusted	Not seasonally adjusted	Seasonally adjusted
Amount outstanding										
31 Dec										
	AEAS		AGAK	AECE		AECK	ATLH		AILA	
1990	312 262		50 999	398 303		73 180	..		52 899	
Net transactions										
	AEAT		AEAZ	AECF		AECL	ATLP		-AIKL	
1987	33 822		6 934	42 452		10 565	6 708		6 017	
1988	36 548		2 818	56 171		9 608	7 541		6 447	
1989	46 470		8 158	63 570		16 236	6 393		6 173	
1990	32 344		11 546	45 122		2 871	4 837		4 130	
		AEAW			AECI			ATLX		-AIKM
1990 Q1	10 668	11 793	4 364	16 788	15 500	45	747	1 161	478	980
Q2	12 388	10 810	2 318	9 906	11 420	154	1 413	996	1 196	1 067
Q3	7 035	4 820	4 567	9 449	7 490	465	1 446	1 398	1 347	1 079
Q4	2 253	4 921	297	8 979	10 713	2 207	1 231	928	1 109	1 005
1991 Q1	740	2 294	462	5 069	3 715	1 913	-111	646	..	..
1990 Jun	8 067	..	573	7 466	..	-96	..	..	..	..
Jul	-583	..	1 700	624	..	941	..	..	..	..
Aug	3 230	..	1 718	1 402	..	547	..	..	..	..
Sep	4 388	..	1 149	7 423	..	-1 023	..	..	..	..
Oct	-633	..	-23	1 844	..	215	..	..	..	..
Nov	4 054	..	768	3 147	..	1 806	..	..	..	..
Dec	-1 168	..	-448	3 988	..	186	..	..	..	..
1991 Jan	-3 928	..	185	-1 251	..	563	..	..	..	..
Feb	811	..	-372	1 472	..	581	..	..	..	..
Mar	3 857	..	649	4 848	..	769	..	..	..	..
Apr	1 681	..	796	220	..	-408	..	..	..	..

1 For further details see *Financial Statistics*, Tables 3.9, 6.1, 6.8, 6.9, 7.1, 7.5, 7.7, 9.3, 12.1.
2 Total administered by the Department for National Savings.
3 Monthly figures relate to calendar months.

Sources: Central Statistical Office;
Department for National Savings;
Building Societies Association;
Unit Trust Association;
Bank of England;
Department of Trade and Industry

From: Monthly Digest of Statistics, June 1991, Table 17.3

Definitions and sources

The object of the balance of payments accounts is to identify and record transactions between residents of the United Kingdom and residents overseas (non-residents) in a way that is suitable for analysing the economic relations between the UK economy and the rest of the world. In the UK balance of payments accounts, transactions are classified into main groups as follows:

Current account transactions cover exports and imports of goods and services, investment income and most transfers.

Transactions in UK external assets and liabilities cover inward and outward investment, transactions by banks in the United Kingdom, borrowing and lending overseas by other UK residents, drawings on and accruals to the official reserves and other capital transactions.

The *current balance* shows whether the United Kingdom has had a surplus of income over expenditure; and, taken with capital transfers, it shows whether the United Kingdom has added to or consumed its net external assets in any period.

In concept every balance of payments transaction involves equal credit and debit entries, relating to the two halves of the transaction so that the accounts are analogous to a double-entry book-keeping system. For example an export of goods, recorded as positive, would be matched by a negative entry, which could be one of the following:

(i)   an increase in the foreign assets (claims on non-residents) of the United Kingdom, (eg an increase in UK residents' deposits with banks abroad) ;

(ii) a decrease in the United Kingdom's liabilities to non-residents (eg a fall in sterling deposits with UK banks) ;

o r (iii) in the case of a barter transaction, by imports of similar value.

Conversely imports of goods, recorded as negative, are likely to be matched by positive entries representing reductions in foreign assets held by the United Kingdom or increases in the United Kingdom's liabilities to non-residents.

Since the two entries made in respect of each transaction are generally derived from separate sources and methods of estimation are neither complete nor precisely accurate, the two entries may not match each other precisely or may fall within different recording periods. In order to bring the total of all entries to zero an additional entry, the balancing item, is therefore included to reflect the sum of all these errors and omissions. The balancing item will include persistent elements, where certain types of transactions are not adequately covered in the accounts.

The balance of payments estimates are compiled from a large number of different sources and the degree of accuracy attained varies considerably between items. Errors are likely, to some extent, to offset each other in any particular year but where a balance is drawn between two aggregates and the balance is small in relation to the aggregates, such as the current balance, the proportionate error attached to the balance is liable to be very substantial.

Detailed notes and definitions relating to the balance of payments (including foreign trade) are given in the annual publication *United Kingdom Balance of Payments* (the CSO Pink Book).

For other sources see:

Guide to Official Statistics, 1990 edition (200 pages approximately fully indexed).

UK Balance of Payments, "CSO Pink Book" .

Overseas Trade Statistics of the United Kingdom.

6.1 Summary balance of payments

£ million

		1980	1981	1982	1983	1984	1985	1986	1987	1988	1989	1990
Current account												
Visible balance	HCHL	1 357	3 251	1 911	-1 537	-5 336	-3 345	-9 559	-11 582	-21 624	-24 598	-18 675
Invisibles												
Services balance	CGIN	3 653	3 792	3 022	4 064	4 519	6 687	6 808	6 745	4 574	4 685	5 201
Interest, profits and dividends balance	CGOA	-182	1 251	1 460	2 831	4 357	2 646	5 096	4 078	5 047	4 088	4 029
Transfers balance	CGIO	-1 984	-1 547	-1 741	-1 593	-1 730	-3 111	-2 157	-3 400	-3 518	-4 578	-4 935
Invisibles balance	CGIK	1 487	3 496	2 741	5 302	7 146	6 222	9 747	7 423	6 103	4 195	4 295
Current balance	AIMG	2 843	6 748	4 649	3 765	1 811	2 878	187	-4 159	-15 520	-20 404	-14 380
Capital transfers	AAAZ	-	-	-	-	-	-	-	-	-	-	-
Transactions in UK assets and liabilities												
UK external assets	HEPZ	-43 439	-50 769	-31 433	-30 378	-31 918	-50 501	-92 663	-79 627	-55 426	-83 199	-72 301
UK external liabilities	HEQW	39 499	43 334	28 916	25 818	24 153	46 419	85 430	85 438	65 071	96 115	84 381
Net transactions	HEQU	-3 940	-7 436	-2 519	-4 562	-7 766	-4 082	-7 234	5 810	9 645	12 916	12 081
EEA loss on forward commitments	HCHF	-	-	-	-	-	-	-	-	-	-	-
Allocation of special drawing rights	HBUN	180	158	-	-	-	-	-	-	-	-	-
Gold subscription to IMF	HBWO	-	-	-	-	-	-	-	-	-	-	-
Balancing item	AASA	917	530	-2 130	797	5 955	1 204	7 047	-1 651	5 875	7 488	2 299

From: United Kingdom Balance of Payments, 1991, Table 1.1

6.2 Current account

£ million

		1980	1981	1982	1983	1984	1985	1986	1987	1988	1989	1990
Credits												
Exports (f.o.b.)	CGJP	47 149	50 668	55 331	60 700	70 265	77 991	72 627	79 153	80 346	92 389	102 038
Services												
General government	CGJR	315	401	404	470	474	483	511	521	551	449	432
Private sector and public corporations												
Sea transport	CGJW	3 789	3 731	3 215	3 043	3 244	3 211	3 216	3 282	3 526	3 870	3 847
Civil aviation	CGJO	2 210	2 359	2 471	2 665	2 931	3 078	2 786	3 159	3 192	3 758	4 358
Travel	CGKA	2 961	2 970	3 188	4 003	4 614	5 442	5 553	6 260	6 184	6 945	7 784
Financial and other services	HHDE	6 192	7 303	8 085	9 175	10 324	12 003	13 626	14 656	14 035	15 380	15 649
Interest, profits and dividends												
General government	CGNR	946	971	979	765	818	735	765	931	1 456	1 948	1 810
Private sector and public corporations	CGNT	22 735	36 559	43 419	41 685	50 800	51 534	46 927	47 138	55 267	72 222	79 477
Transfers												
General government	HDKD	958	1 675	2 154	2 235	2 392	1 760	2 138	2 282	2 115	2 143	2 193
Private sector	CGJV	935	1 117	1 248	1 528	1 652	1 775	1 732	1 666	1 715	1 750	1 800
Total invisibles	CGJY	41 041	57 085	65 162	65 569	77 249	80 022	77 253	79 896	88 041	108 465	117 350
Total credits	CGPZ	88 190	107 753	120 493	126 269	147 514	158 013	149 880	159 049	168 387	200 854	219 388
Debits												
Imports (f.o.b.)	CGGL	45 792	47 416	53 421	62 237	75 601	81 336	82 186	90 735	101 970	116 987	120 713
Services												
General government	CGGI	1 165	1 264	1 754	1 522	1 655	1 781	1 920	2 141	2 351	2 698	2 753
Private sector and public corporations												
Sea transport	CGGW	3 739	3 818	3 589	3 665	3 600	3 508	3 302	3 310	3 566	3 737	3 591
Civil aviation	CGGG	1 863	2 005	2 184	2 363	2 676	2 877	3 194	3 775	4 097	4 298	4 674
Travel	CGHA	2 738	3 272	3 640	4 090	4 663	4 871	6 083	7 280	8 216	9 357	9 916
Financial and other services	HBVH	2 309	2 613	3 174	3 652	4 474	4 493	4 385	4 627	4 684	5 627	5 935
Interest, profits and dividends												
General government	HERS	895	940	1 090	1 188	1 329	1 479	1 668	2 037	2 315	2 500	2 400
Private sector and public corporations	HHII	22 970	35 338	41 847	38 429	45 931	48 145	40 929	41 955	49 361	67 583	74 858
Transfers												
General government	CGGJ	2 738	3 282	3 943	4 165	4 491	5 187	4 371	5 559	5 363	6 421	6 828
Private sector	CGGV	1 139	1 057	1 200	1 191	1 283	1 459	1 656	1 789	1 985	2 050	2 100
Total invisibles	CGGY	39 555	53 589	62 423	60 267	70 102	73 799	67 507	72 473	81 937	104 271	113 055
Total debits	CGQB	85 347	101 005	115 844	122 504	145 703	155 135	149 693	163 208	183 907	221 258	233 768
Balances												
Visible balance	HCHL	1 357	3 251	1 911	-1 537	-5 336	-3 345	-9 559	-11 582	-21 624	-24 598	-18 675
Services												
General government	CGIG	-850	-863	-1 350	-1 052	-1 181	-1 298	-1 409	-1 620	-1 800	-2 249	-2 321
Private sector and public corporations												
Sea transport	HBTO	50	-87	-374	-622	-356	-297	-86	-28	-40	133	256
Civil aviation	HDJA	347	354	287	302	255	201	-408	-616	-905	-540	-316
Travel	HBYE	223	-302	-452	-87	-49	571	-530	-1 020	-2 032	-2 412	-2 132
Financial and other services	HHCW	3 883	4 690	4 911	5 523	5 850	7 510	9 241	10 029	9 351	9 753	9 714
Interest, profits and dividends												
General government	HERV	52	30	-112	-424	-511	-744	-902	-1 105	-859	-552	-591
Private sector and public corporations	CGQD	-235	1 221	1 571	3 255	4 870	3 387	5 998	5 183	5 906	4 639	4 619
Transfers												
General government	HDKH	-1 780	-1 607	-1 789	-1 930	-2 099	-3 427	-2 233	-3 277	-3 248	-4 278	-4 635
Private sector	CGIM	-204	60	48	337	369	316	76	-123	-270	-300	-300
Invisibles balance	CGIK	1 487	3 496	2 741	5 302	7 146	6 222	9 747	7 423	6 103	4 195	4 295
Of which:Private sector and public corporations:Services and IPD	HBZC	4 268	5 876	5 943	8 371	10 570	11 372	14 215	13 548	12 280	11 573	12 141
Current balance	AIMG	2 843	6 748	4 649	3 765	1 811	2 878	187	-4 159	-15 520	-20 404	-14 380

From: United Kingdom Balance of Payments, 1991, Table 1.3

BALANCE OF PAYMENTS

6.3 Summary of transactions in UK external assets and liabilities
(Capital account of the United Kingdom with the overseas sector)

£ million

		1980	1981	1982	1983	1984	1985	1986	1987	1988	1989	1990
Transactions in external assets of the UK (increase in assets shown negative)												
Direct investment overseas by UK residents	HHBV	-4 867	-6 005	-4 091	-5 417	-6 033	-8 456	-12 038	-19 215	-20 880	-21 521	-11 702
Portfolio investment in overseas securities by UK residents	CGOS	-3 310	-4 467	-7 565	-7 350	-9 759	-16 755	-22 095	7 201	-8 600	-31 283	-12 587
Lending etc to overseas residents by UK banks [1]	HEYN	-32 614	-39 919	-20 566	-18 443	-14 359	-22 024	-53 678	-50 427	-19 515	-27 032	-37 246
Deposits and lending overseas by UK residents other than banks and general government												
Transactions with banks abroad	HESZ	-2 502	-1 864	-598	863	-3 213	-1 305	-3 109	-4 632	-3 980	-9 473	-5 722
Other assets	HETE	-209	-161	126	-161	1 281	528	1 656	254	1 201	1 611	-3 740
Official reserves	AIPA	-291	2 419	1 421	607	908	-1 758	-2 891	-12 012	-2 761	5 440	-77
Other external assets of central government	HEUJ	351	93	-161	-478	-743	-730	-509	-796	-891	-942	-1 227
Total transactions in assets of												
General government	HCDN	60	2 512	1 261	129	165	-2 488	-3 401	-12 808	-3 652	4 498	-1 304
Public corporations	HEYO	76	-167	-210	47	-223	370	-121	-21	-33	-59	-48
UK banks	HCDG HEYP	-43 576	-41 133	-22 550	-21 232	-21 402	-33 347	-63 068	-49 484	-20 738	-34 285	-43 826
UK non-bank private sector	HEYQ		-11 981	-9 934	-9 322	-10 457	-15 035	-26 073	-17 313	-31 003	-53 354	-27 122
Total	HEPZ	-43 439	-50 769	-31 433	-30 378	-31 918	-50 501	-92 663	-79 627	-55 426	-83 199	-72 301
Transactions in UK liabilities to overseas residents (increase in liabilities shown positive)												
Direct investment in the UK by overseas residents	HHBU	4 355	2 932	3 027	3 386	-181	3 865	4 987	8 478	10 236	17 145	18 997
Portfolio investment in the UK by overseas residents	HEYR	1 431	257	-11	1 701	1 288	9 671	11 785	19 210	14 387	13 239	5 070
Borrowing etc from overseas residents by UK banks [1]	HEYS	33 549	39 260	24 421	21 293	24 790	29 443	64 127	52 600	34 218	43 887	46 179
Borrowing from overseas by UK residents other than banks and general government												
Transactions with banks abroad	HETN	457	864	950	38	-2 263	2 682	3 787	1 870	3 718	5 949	7 789
Other liabilities	HETQ	301	224	119	-15	558	732	567	1 448	1 682	13 710	5 649
Other external liabilities of general government	HEUR	-594	-206	409	-584	-40	24	178	1 829	831	2 186	697
Total transactions in liabilities of												
General government	HEYT	877	164	643	347	584	3 235	3 521	5 834	2 240	240	-3 847
Public corporations	HEYU	-269	-567	-203	-46	-260	-51	-28	-247	-277	-1 481	-84
UK private sector	HEYV	38 890	43 736	28 476	25 517	23 830	43 235	81 936	79 849	63 109	97 356	88 312
Total	HEQW	39 499	43 334	28 916	25 818	24 153	46 419	85 430	85 438	65 071	96 115	84 381
Of which : identified liabilities constituting overseas authorities' exchange reserves in sterling	HCRC	1 262	89	226	981	1 308	1 577	36	4 494	2 413	904	1 465
Net transactions in UK external assets and liabilities												
Direct investment	HHPD	-512	-3 073	-1 064	-2 031	-6 214	-4 591	-7 051	-10 737	-10 644	-4 376	7 295
Portfolio investment	HHPH	-1 878	-4 210	-7 575	-5 649	-8 472	-7 084	-10 310	26 411	5 787	-18 044	-7 517
Lending and borrowing by UK banks	HCAI	935	-659	3 855	2 850	10 431	7 419	10 449	2 173	14 703	16 855	8 933
Lending and borrowing by UK residents other than banks and general government	HETZ	-1 954	-1 800	598	726	-3 636	2 636	2 900	-1 059	2 620	11 796	3 976
Other external assets and liabilities of general government	HCCG	-533	2 307	1 668	-457	124	-2 464	-3 223	-10 979	-2 822	6 684	-607
Net transactions in assets and liabilities of												
General government	HEYW	938	2 677	1 903	474	747	747	120	-6 974	-1 413	4 738	-5 151
Public corporations	HEYX	-192	-734	-413	-	-484	319	-150	-268	-311	-1 540	-132
UK private sector	HEYY	-4 686	-9 379	-4 007	-5 037	-8 030	-5 149	-7 204	13 053	11 367	9 718	17 364
Total net	HEQU	-3 940	-7 436	-2 519	-4 562	-7 766	-4 082	-7 234	5 810	9 645	12 916	12 081
Allocations of Special Drawing Rights to the UK by the IMF	HBUN	180	158	-	-	-	-	-	-	-	-	-

From: United Kingdom Balance of Payments, 1991, Table 7.1

6.4 UK external assets by sector[1]

£ million

		1980	1981	1982	1983	1984	1985	1986	1987	1988	1989	1990
Net												
General government	HEVL	8 722	7 511	7 032	6 938	7 616	5 324	6 549	10 912	11 529	8 396	10 396
Public corporations	HEVM	-3 395	-3 216	-3 330	-3 586	-3 533	-3 163	-3 188	-2 261	-2 025	-7	132
Private sector	HEVN	12 769	28 348	39 788	52 212	75 314	71 673	100 281	58 282	71 465	75 659	19 315
Total	HEVO	18 095	32 643	43 490	55 565	79 397	73 834	103 642	66 933	80 970	84 048	29 843

[1] Because of the many inconsistencies in valuing the component series, and the omission of certain assets and liabilities which are unidentifiable, these estimates are not an exact measure of the UK's external debtor/ creditor position.

From: United Kingdom Balance of Payments, 1991, Table 1.2

FOREIGN TRADE

Definitions and sources

Detailed figures, notes and definitions relating to foreign trade are given in the annual and monthly Overseas Trade Statistics of the United Kingdom.

Data on unit value, and volume indices are given in the Monthly Review of External Trade Statistics.

For other sources see:

Guide to Official Statistics, 1990 edition (200 pages approximately fully indexed) HMSO.

7.1 Value of Exports (f.o.b) and Imports (c.i.f): analysis by areas

£ million, seasonally adjusted

	European Community	Rest of Western Europe	E Europe & Soviet Union	North America	Other OECD	Oil exporting countries	Other Countries	Total
Exports								
	BOGB	BOGC	OBWN	BOGD	BOGE	BOGF	OBWE	CGKI
1985	38 223	7 410	1 299	13 310	3 789	5 950	9 325	78 263
1986	35 025	6 918	1 275	12 065	3 602	5 495	9 000	72 812
1987	39 497	7 415	1 241	12 992	3 195	5 220	9 786	79 849
1988	41 052	7 210	1 285	12 794	3 520	5 019	10 009	82 098
1989	47 540	7 987	1 473	14 437	4 519	5 831	11 084	93 771
1990	55 071	9 041	1 480	14 973	4 824	5 575	12 171	103 882
1989 Oct	4 138	795	134	1 269	384	490	958	8 173
Nov	4 188	706	152	1 236	446	495	1 010	8 114
Dec	4 534	705	140	1 347	435	552	1 098	8 921
1990 Jan	4 477	738	131	1 245	408	443	1 056	8 600
Feb	4 322	742	146	1 300	392	466	1 011	8 526
Mar	4 353	751	135	1 296	381	446	1 011	8 491
Apr	4 590	821	129	1 258	364	488	1 034	8 622
May	4 609	822	125	1 255	475	617	1 031	8 902
Jun	4 488	800	108	1 270	378	526	1 001	8 645
Jul	4 434	719	124	1 236	384	390	987	8 307
Aug	4 521	675	120	1 178	379	496	1 029	8 610
Sep	4 753	707	126	1 291	419	505	1 025	9 013
Oct	4 806	792	126	1 199	436	433	1 007	8 798
Nov	4 932	778	119	1 295	424	427	981	8 823
Dec	4 786	696	91	1 150	384	338	998	8 545
1991 Jan	4 874	716	84	906	328	463	969	8 377
Feb	4 823	674	102	997	300	374	938	8 389
Mar	4 840	685	113	1 014	341	446	942	8 516
Apr	4 917	754	106	1 120	312	469	991	8 554
May	4 955	732	103	1 122	343	472	961	8 612
Imports								
	BOGJ	BOGK	OBWQ	BOGL	BOGM	BOGN	OBWH	CGHM
1985	41 650	12 075	1 496	11 697	6 352	2 810	9 790	84 904
1986	44 727	11 863	1 477	9 994	6 839	2 062	8 931	85 663
1987	49 736	12 710	1 696	10 781	6 722	1 699	10 243	94 026
1988	55 958	13 831	1 629	12 903	7 817	2 085	11 663	106 572
1989	63 807	15 155	1 781	15 929	8 514	2 313	13 659	121 888
1990	65 955	15 715	1 797	16 751	8 414	2 974	13 748	126 135
1989 Oct	5 408	1 347	144	1 281	726	170	1 125	10 285
Nov	5 474	1 291	144	1 403	709	210	1 179	10 456
Dec	5 489	1 118	130	1 316	719	238	1 139	10 225
1990 Jan	5 635	1 475	146	1 489	736	339	1 180	11 033
Feb	5 491	1 287	138	1 450	666	250	1 049	10 431
Mar	5 796	1 416	151	1 531	782	273	1 097	11 127
Apr	5 748	1 347	152	1 524	746	185	1 171	11 007
May	5 698	1 357	144	1 508	767	241	1 177	10 920
Jun	5 525	1 241	174	1 520	706	190	1 276	10 671
Jul	5 483	1 253	141	1 433	685	239	1 172	10 450
Aug	5 449	1 187	138	1 252	663	201	1 132	10 222
Sep	5 499	1 238	142	1 152	635	304	1 090	10 054
Oct	5 333	1 378	157	1 264	683	238	1 165	10 286
Nov	5 235	1 319	165	1 290	712	309	1 107	10 149
Dec	5 063	1 217	149	1 338	633	205	1 132	9 785
1991 Jan	5 142	1 417	189	1 283	652	263	1 096	9 982
Feb	5 124	1 048	129	1 197	582	169	1 065	9 461
Mar	5 171	1 156	133	1 199	722	207	1 185	9 728
Apr	4 913	1 207	116	1 336	715	208	1 193	9 733
May	5 086	1 222	121	1 355	662	205	1 252	9 861

The statistics are on an overseas trade statistics basis (see footnote 1 to Table 15.7).

Source: Central Statistical Office

From: Monthly Digest of Statistics, June 1991, Table 15.4

FOREIGN TRADE

7.2 Value of Exports (f.o.b) and Imports (c.i.f): analysis by commodity classes

£ million, seasonally adjusted

	Total	Food, beverages and tobacco	Basic materials	Fuels	Total manufactures	Manufactures excluding erratics[1] Semi-manufactures[2] Chemicals	Other	Total	Finished manufactures[3] Passenger motor cars[4]	Other consumer[4]	Inter-mediate[4]	Capital[4]	Total	Total
SITC (Rev. 3) Section, Division or Group		0 and 1	2 and 4	3	5 to 8	5	6 less PS	5 and 6 less PS					7 and 8 less SNA	5 to 8 less SNAPS
Exports														
	CGKI	BOCB	BOCC	BOCD	BOCE	BOCH	BOCI	BOCG	BOCK	BOCL	BOCM	BOCN	BOCJ	BOCF
1985	78 263	4 976	2 183	16 776	52 474	9 431	8 896	18 328	1 343	5 057	14 119	9 595	30 113	48 441
1986	72 812	5 490	2 100	8 653	54 546	9 696	8 936	18 632	1 363	5 530	14 320	9 803	31 016	49 648
1987	79 849	5 599	2 243	8 746	61 005	10 541	9 817	20 359	1 981	6 788	15 250	11 036	35 053	55 409
1988	82 098	5 534	2 120	6 258	66 194	11 331	10 578	21 910	2 033	6 644	16 012	13 367	38 056	59 967
1989	93 771	6 555	2 349	6 175	76 407	12 351	12 211	24 561	2 639	8 029	18 237	15 651	44 555	69 117
1990	103 882	7 094	2 251	7 774	84 469	13 182	13 552	26 735	3 324	9 565	20 618	16 808	50 317	77 050
1989 Oct	8 173	529	210	605	6 596	1 061	1 078	2 140	230	754	1 585	1 366	3 934	6 074
Nov	8 114	533	195	622	6 562	1 063	1 013	2 076	243	605	1 569	1 392	3 809	5 885
Dec	8 921	603	203	594	7 319	1 153	1 140	2 293	248	755	1 646	1 474	4 122	6 415
1990 Jan	8 600	602	183	692	6 923	1 145	1 120	2 265	229	790	1 597	1 414	4 031	6 296
Feb	8 526	563	185	639	6 929	1 074	1 125	2 199	251	770	1 654	1 441	4 116	6 315
Mar	8 491	545	177	661	6 922	1 095	1 149	2 244	236	769	1 659	1 459	4 123	6 367
Apr	8 622	545	186	612	7 067	1 115	1 141	2 256	243	725	1 695	1 503	4 166	6 422
May	8 902	541	200	576	7 392	1 139	1 150	2 290	266	873	1 746	1 549	4 434	6 724
Jun	8 645	593	185	530	7 150	1 114	1 133	2 247	275	908	1 741	1 422	4 346	6 592
Jul	8 307	591	181	444	6 916	1 102	1 134	2 236	225	758	1 787	1 322	4 093	6 329
Aug	8 610	661	198	594	6 955	1 037	1 097	2 134	296	824	1 710	1 301	4 131	6 265
Sep	9 013	622	199	815	7 180	1 124	1 153	2 277	292	850	1 806	1 350	4 298	6 575
Oct	8 798	616	199	785	7 022	1 075	1 127	2 202	302	802	1 735	1 359	4 198	6 400
Nov	8 823	598	187	738	7 116	1 104	1 134	2 238	357	764	1 759	1 364	4 244	6 482
Dec	8 545	617	171	688	6 897	1 058	1 089	2 147	352	732	1 729	1 324	4 137	6 283
1991 Jan	8 377	642	166	491	6 902	1 092	1 090	2 182	347	761	1 680	1 305	4 093	6 275
Feb	8 389	625	170	567	6 869	1 078	1 088	2 166	340	750	1 707	1 293	4 090	6 255
Mar	8 516	600	162	609	6 985	1 105	1 113	2 218	319	791	1 769	1 365	4 245	6 463
Apr	8 554	595	171	494	7 151	1 167	1 147	2 314	338	785	1 856	1 358	4 338	6 652
May	8 612	623	171	545	7 136	1 111	1 124	2 234	344	782	1 810	1 395	4 331	6 565
Imports														
	CGHM	BODB	BODC	BODD	BODE	BODH	BODI	BODG	BODK	BODL	BODM	BODN	BODJ	BODF
1985	84 904	9 275	5 496	10 648	58 286	6 916	12 578	19 494	4 165	8 970	12 141	10 137	35 414	54 908
1986	85 663	10 031	5 066	6 384	62 800	7 360	13 234	20 595	4 809	10 144	13 248	10 642	38 843	59 438
1987	94 026	10 130	5 688	6 099	70 967	8 347	14 970	23 316	5 024	11 488	15 365	12 181	44 060	67 374
1988	106 572	10 616	5 983	5 039	83 484	9 314	17 423	26 737	6 750	12 603	17 915	14 616	51 885	78 623
1989	121 888	11 429	6 491	6 429	95 974	10 439	19 530	29 969	7 619	14 827	21 199	16 818	60 462	90 429
1990	126 135	12 316	6 098	7 809	98 191	10 834	19 707	30 543	7 398	15 809	22 251	15 885	61 349	91 888
1989 Oct	10 285	1 023	505	572	8 057	892	1 620	2 512	654	1 270	1 790	1 423	5 136	7 648
Nov	10 456	998	558	601	8 133	901	1 640	2 541	641	1 233	1 840	1 389	5 103	7 643
Dec	10 225	985	472	565	8 098	886	1 617	2 503	652	1 318	1 852	1 395	5 217	7 720
1990 Jan	11 033	1 100	563	678	8 550	896	1 641	2 537	640	1 339	1 930	1 443	5 352	7 888
Feb	10 431	1 004	516	648	8 126	823	1 561	2 385	654	1 315	1 887	1 385	5 241	7 626
Mar	11 127	1 074	545	631	8 724	916	1 697	2 613	639	1 272	1 888	1 428	5 227	7 840
Apr	11 007	1 060	578	568	8 654	931	1 714	2 645	640	1 444	1 949	1 430	5 464	8 109
May	10 920	1 073	552	589	8 561	921	1 687	2 608	659	1 383	1 940	1 445	5 427	8 035
Jun	10 671	1 045	547	507	8 419	924	1 674	2 598	625	1 368	1 907	1 352	5 252	7 850
Jul	10 450	1 031	524	563	8 207	901	1 690	2 591	717	1 313	1 811	1 288	5 130	7 721
Aug	10 222	1 013	486	551	8 030	903	1 648	2 551	606	1 331	1 822	1 276	5 037	7 587
Sep	10 054	998	473	675	7 757	881	1 644	2 525	579	1 276	1 804	1 250	4 909	7 434
Oct	10 286	992	450	887	7 814	889	1 616	2 505	596	1 261	1 753	1 206	4 817	7 321
Nov	10 149	987	441	774	7 808	919	1 621	2 540	530	1 263	1 793	1 211	4 798	7 338
Dec	9 785	939	423	738	7 541	930	1 514	2 445	513	1 244	1 767	1 171	4 695	7 139
1991 Jan	9 982	972	406	758	7 707	935	1 574	2 509	479	1 243	1 740	1 163	4 624	7 133
Feb	9 461	983	413	572	7 349	899	1 526	2 426	483	1 174	1 774	1 155	4 585	7 011
Mar	9 728	1 017	412	578	7 603	877	1 551	2 429	452	1 249	1 809	1 204	4 716	7 144
Apr	9 733	1 012	408	603	7 545	826	1 524	2 350	461	1 265	1 775	1 209	4 710	7 060
May	9 861	1 008	425	616	7 662	882	1 525	2 407	462	1 282	1 841	1 242	4 828	7 235

The statistics are on an overseas trade basis (see footnote 1 to Table 15.7)
1 These are defined as ships, North Sea installations (together comprising SITC(Rev 3)(793), aircraft (792), precious stones (667) and silver (681.1).
2 Excluding precious stones and silver (PS).
3 Excluding ships, North Sea installations and aircraft (SNA).
4 Based on the Classification by Broad Economic categories (BEC) published by the United Nations.

Source: Central Statistical Office

From: Monthly Digest of Statistics, June 1991, Table 15.1

7.3 Export and Import volume indices

1985 = 100, Seasonally adjusted

	Total	Food, beverages and tobacco	Basic materials	Fuels	Total manufactures	Manufactures excluding erratics[1]								
							Semi-manufactures excluding precious stones (P) and silver			Finished manufactures exc ships, North Sea installations and aircraft SNA				
						Total	Total	Chemicals	Other	Total	Passenger motor cars[2]	Other consumer[2]	Intermediate[2]	Capital[2]
SITC (Rev 3) Section or division	0 to 9	0 and 1	2 and 4	3	5 to 8	5 to 8 less SNAPS	5 and 6 less PS	5	6 less PS	7 and 8 less SNA				
Weights	1 000	64	28	214	670	619	234	121	114	385	17	65	180	122
Exports														
	BOKO	BOKP	BOKQ	BOKR	BOKS	BOKT	BOKU	BOKV	BOKW	BOKX	BOKY	BOKZ	BOKA	BOKB
1986	104.0	108.2	106.4	104.1	103.5	101.7	102.1	103.8	100.3	101.4	92.5	105.6	100.5	101.7
1987	109.8	112.1	114.4	100.8	112.2	110.4	110.0	111.2	108.6	110.6	117.3	125.7	103.5	112.3
1988	113.0	112.3	99.9	93.8	120.3	117.6	116.5	117.3	115.7	118.2	122.8	122.5	104.5	135.4
1989	118.0	123.5	104.3	75.1	132.1	128.9	121.1	118.3	124.1	133.6	153.1	144.4	114.5	153.1
1990	126.0	124.1	102.0	80.7	142.4	140.0	128.1	121.3	135.3	147.3	183.4	169.1	126.8	160.9
1990 Feb	125.0	118	100	84	140	137	125	118	133	145	165	161	122	167
Mar	125.5	112	92	91	140	139	129	122	136	145	161	161	122	168
Apr	126.7	111	97	87	142	140	128	121	135	147	158	152	124	176
May	129.4	111	105	82	148	145	130	122	138	154	170	180	128	178
Jun	126.6	122	98	82	143	143	127	121	133	152	176	194	128	162
Jul	120.0	122	97	69	138	136	126	120	133	142	140	159	131	151
Aug	125.3	140	110	75	141	138	124	115	134	146	191	176	128	150
Sep	129.1	132	111	83	145	144	132	125	139	151	196	182	133	155
Oct	126.7	131	114	79	143	141	130	122	139	148	215	171	128	155
Nov	127.4	128	108	76	146	142	131	124	139	149	241	167	130	155
Dec	124.5	134	98	73	142	139	125	118	133	147	239	160	130	152
1991 Jan	120.7	132	97	57	142	139	129	121	136	145	230	168	126	149
Feb	125.9	127	103	79	143	139	128	121	137	146	227	164	129	150
Mar	126.7	120	96	83	144	144	132	125	140	151	209	173	131	160
Apr	124.8	118	99	68	146	147	137	129	146	154	226	171	138	159
May	124.5	124	98	71	144	144	132	123	142	151	224	168	130	162
Weights	1 000	109	65	125	686	647	230	81	148	417	49	106	143	119
Imports														
	BONO	BONP	BONQ	BONR	BONS	BONT	BONU	BONV	BONW	BONX	BONY	BONZ	BONA	BONB
1986	107.2	109.0	107.3	106.9	106.8	107.1	106.3	108.3	105.2	107.6	102.7	112.4	109.5	103.0
1987	114.8	109.4	116.9	105.1	117.8	118.6	116.9	119.7	115.3	119.5	99.1	123.7	118.1	115.8
1988	131.2	115.0	118.1	108.3	139.5	139.2	131.8	129.6	133.1	143.3	130.1	141.6	148.1	144.4
1989	141.6	118.3	116.7	117.7	152.5	152.3	140.2	142.6	138.8	159.0	135.1	159.0	164.9	161.6
1990	143.8	121.5	114.3	125.4	153.9	153.3	145.1	148.6	143.1	157.8	121.4	169.1	166.6	152.1
1990 Feb	140.3	118	109	125	150	149	133	136	132	158	132	161	168	155
Mar	150.3	125	120	134	161	154	146	152	143	159	128	158	167	161
Apr	148.6	123	121	127	160	159	147	154	143	166	125	179	172	162
May	147.0	122	121	131	157	157	145	149	143	163	125	172	171	160
Jun	144.5	124	123	117	155	155	145	148	144	160	121	173	169	154
Jul	145.4	122	117	139	154	154	149	150	149	157	140	168	162	149
Aug	142.1	121	108	119	153	154	148	151	146	157	119	176	164	148
Sep	139.8	119	110	121	150	152	146	147	145	156	117	170	164	149
Oct	143.3	122	108	131	153	152	150	148	151	153	117	171	161	144
Nov	140.6	119	112	117	151	151	148	150	147	153	106	171	166	142
Dec	134.6	115	106	114	144	146	141	151	136	149	98	166	162	138
1991 Jan	141.4	120	107	129	151	147	147	155	143	147	93	167	161	136
Feb	133.0	118	105	112	142	144	141	147	138	146	94	156	164	135
Mar	139.6	124	114	122	148	149	145	145	145	151	87	166	169	141
Apr	140.2	126	109	133	147	146	141	136	144	148	85	164	164	141
May	138.1	119	108	126	146	147	140	148	136	150	84	164	167	145

The statistics are on an overseas trade statistics basis (see footnote 1 to Table 15.7).

1 These are defined as ships, North Sea installations (together comprising SITC (Rev 3) (793), aircraft (792), precious stones (667) and silver (6811).

2 Based on the *Classification by Broad Economic Categories*, (BEC) published by the United Nations.

Source: Central Statistical Office

From: Monthly Digest of Statistics, June 1991, Table 15.8

Definitions and sources

The General Index of Retail Prices (RPI) measures the changes month by month in the level of prices of the goods and services purchased by all types of household in the UK, with the exception of some higher income households and retired people mainly dependent on state benefits. A special pensioner price index is published in *Economic Trends* and in the *Department of Employment Gazette*.

The Tax and Price Index (TPI) is described in a footnote to the table in which it appears.

The Producer Price Index has replaced what was called the Wholesale Price Index. The new title is more accurate and conforms with international nomenclatures. The index was also re-based from 1975 to 1980. Full details were given in an article in *British Business*, 15 April 1983.

For other sources see:

Guide to Official Statistics, 1990 edition (200 pages approximately fully indexed) HMSO.

8.1 General index of retail prices[1]

	All items	All items except seasonal food[2]	Food	Alcoholic drink	Tobacco	Housing	Fuel and light	Durable household goods	Clothing and footwear	Transport and vehicles	Miscella neous goods	Services	Meals bought and consumed outside the home
15 January 1974=100													
Annual averages													
	CBAB	CBAP	CBAN	CBAA	CBAC	CBAH	CBAG	CBAE	CBAD	CBAO	CBAJ	CBAM	CBAI
1982	320.4	322.0	299.3	341.0	413.3	358.3	433.3	243.8	210.5	343.5	325.8	331.6	341.7
1983	335.1	337.1	308.8	366.4	440.9	367.1	465.4	250.4	214.8	366.3	345.6	342.9	364.0
1984	351.8	353.1	326.1	387.7	489.0	400.7	478.8	256.7	214.6	374.7	364.7	357.3	390.8
1985	373.2	375.4	336.3	412.1	532.4	452.3	499.3	263.9	222.9	392.5	392.2	381.3	413.3
1986	385.9	387.9	347.3	430.6	584.9	478.1	506.0	266.7	229.2	390.1	409.2	400.5	439.5
1987 Jan 1	394.5	396.4	354.0	100.0	602.9	502.4	506.1	265.6	230.8	399.7	413.0	408.8	454.8

	All items	Food and catering	Alcohol and tobacco	Housing and household expend- iture	Personal expend- iture	Travel and leisure	All items except seasonal food[2]	All items except food	Seasonal food[2,3]	Non- seasonal food[3]	All items except housing	National ised industri es[4]	Consumer durables
13 January 1987=100													
Weights 1991	1000	198	109	353	101	239	976	849	24	127	808		128
Annual averages													
	CHAW	CHBS	CHBT	CHBU	CHBV	CHBW	CHAX	CHAY	CHBP	CHBB	CHAZ	CHBX	CHBY
1987	101.9	101.4	101.2	102.1	101.4	102.6	101.9	102.0	101.6	101.0	101.6	100.9	101.2
1988	106.9	105.7	105.7	108.4	105.2	107.2	107.0	107.3	102.4	105.0	105.8	106.7	103.7
1989	115.2	111.9	110.8	121.9	111.2	112.8	115.5	116.1	105.0	111.6	111.5		107.2
1988 Nov 15	110.0	107.1	107.8	114.1	108.1	109.1	110.3	110.9	98.8	107.0	107.8	109.3	105.7
Dec 13	110.3	107.8	107.7	114.3	108.3	109.0	110.5	111.0	101.5	107.4	108.0	109.3	105.9
1989 Jan 17	111.0	108.7	108.5	115.4	107.4	109.8	111.2	111.7	103.2	108.2	108.5	110.9	104.5
Feb 14	111.8	109.0	108.9	116.8	108.4	110.2	111.9	112.5	103.4	108.5	109.0	110.9	105.3
Mar 14	112.3	109.6	109.2	117.4	108.9	110.7	112.4	113.0	104.8	108.9	109.4	110.9	105.8
Apr 18	114.3	110.8	109.6	120.7	111.0	112.3	114.4	115.2	108.0	109.9	110.6	114.2	107.0
May 16	115.0	111.5	109.9	121.4	111.6	113.3	115.1	115.9	109.9	110.4	111.3	114.7	107.5
Jun 13	115.4	111.9	110.2	122.0	111.7	113.6	115.6	116.3	109.3	111.0	111.6	115.9	107.6
Jul 18	115.5	111.6	110.6	122.7	110.7	113.7	115.9	116.6	100.6	111.9	111.6	116.5	106.5
Aug 15	115.8	112.1	111.3	123.2	110.9	113.4	116.2	116.9	100.8	112.3	111.8	116.8	106.7
Sep 12	116.6	112.8	112.0	123.9	111.5	113.9	117.0	117.6	100.7	113.2	112.5	116.9	107.9
Oct 17	117.5	113.8	113.0	124.9	113.7	114.4	117.9	118.5	101.5	114.4	113.3	117.2	108.8
Nov 14	118.5	114.8	113.1	127.2	114.3	114.5	118.9	119.5	106.2	114.8	113.8	117.4	109.3
Dec 16	118.8	115.8	113.1	127.7	114.6	114.0	119.0	119.7	111.1	115.1	114.0	..	109.5
1990 Jan 16	119.5	117.2	113.7	128.4	113.4	114.8	119.6	120.2	116.3	116.0	114.6	..	108.0
Feb 13	120.2	118.1	114.3	129.0	114.7	115.4	120.3	120.9	118.7	116.7	115.3	..	109.1
Mar 13	121.4	118.7	114.8	131.3	115.6	115.9	121.4	122.1	119.6	117.3	115.9	..	109.9
Apr 10	125.1	119.9	118.6	138.5	117.0	118.0	125.1	126.3	123.4	118.0	117.6	..	111.0
May 15	126.2	121.2	120.9	139.8	117.6	118.6	126.3	127.4	123.6	119.4	118.8	..	111.6
Jun 12	126.7	121.3	121.3	140.7	117.5	119.1	126.9	128.0	118.3	120.3	119.1	..	111.5
Jul 17	126.8	120.6	122.4	141.4	116.0	119.6	127.3	128.4	108.1	120.7	119.1	..	109.7
Aug 14	128.1	121.7	123.0	142.5	117.2	121.4	128.5	129.6	112.2	121.4	120.3	..	110.7
Sep 11	129.3	120.3	123.5	143.6	119.3	123.5	129.8	131.1	111.5	121.8	121.6	..	112.5
Oct 16	130.3	122.5	124.4	144.8	120.3	124.6	130.7	132.2	111.8	121.9	122.6	..	113.2
Nov 13	130.0	123.4	124.6	143.8	121.1	123.7	130.4	131.7	114.5	122.4	122.7	..	113.8
Dec 11	129.9	124.1	125.1	143.8	121.1	122.4	130.2	131.4	119.2	122.6	122.6	..	114.1
1991 Jan 15	130.2	124.9	126.0	144.2	118.6	122.8	130.4	131.6	121.2	123.1	122.7	..	110.7
Feb 12	130.9	126.2	126.8	145.0	119.7	123.1	131.1	132.2	125.9	124.0	123.5	..	111.8
Mar 12	131.4	126.4	127.3	145.5	120.9	123.6	131.6	132.8	124.4	124.4	123.9	..	113.0
Apr 16	133.1	128.5	136.9	141.7	123.8	127.5	133.3	134.5	122.5	125.8	127.6	..	115.2
May 14	133.5	128.6	137.9	141.5	124.2	128.9	133.8	135.1	122.5	126.2	128.5	..	116.0

1 Following the recommendation of the Retail Price Index Advisory Committee, the index has been re-referenced to make 13 January, 1987=100. Further details can be found in the April 1987 edition of *Employment Gazette*.

2 Seasonal food is defined as; items of food the prices of which show significant seasonal variations. These are fresh fruit and vegetables, fresh fish, eggs and home-killed lamb.

3 For the February, March and April 1988 indices, the weights for seasonal and non-seasonal food were 24 and 139 respectively. Thereafter the weight for home-killed lamb (a seasonal item) was increased by 1 and that for imported lamb (a non-seasonal item) correspondingly reduced by 1 in the light of new information about their relative shares of household expenditure.

4 "From December 1989 the Nationalised Industries Index is no longer published. Industries remaining nationalised in December 1989 were, Coal, Electricity, Postage and Rail."

Source: Central Statistical Office
From: Monthly Digest of Statistics, June 1991, Table 18.1

8.2 Tax and price index

	January 1978=100										January 1987=100			
	BSAA										DQAB			
	1978	1979	1980	1981	1982	1983	1984	1985	1986	1987	1987	1989	1990	1991
January	100.0	106.1	123.2	140.4	162.3	170.7	177.9	184.7	192.9	198.0	100.0	107.1	113.9	123.6
February	100.7	107.2	125.3	141.9	162.4	171.6	178.8	186.4	193.7	..	100.5	108.0	114.7	124.3
March	101.5	108.2	127.2	144.3	164.0	171.9	179.4	188.4	194.0	..	100.7	108.5	115.9	124.9
April	98.4	110.5	130.8	151.3	166.0	171.8	178.8	190.2	192.5	..	99.7	109.8	118.2	125.4
May	99.1	111.6	132.2	152.4	167.4	172.6	179.6	191.2	192.9	..	99.8	110.5	119.4	125.8
June	100.0	113.8	133.6	153.5	168.0	173.1	180.1	191.7	192.8	..	99.8	110.9	119.9	..
July	100.5	113.8	134.9	154.2	169.0	174.2	179.9	191.3	192.1	..	99.7	111.1	120.0	..
August	101.3	114.9	135.3	155.5	169.0	175.1	181.8	191.8	192.9	..	100.0	111.4	121.4	..
September	101.8	116.2	136.3	156.6	168.9	176.0	183.5	191.4	194.3	..	100.4	112.2	122.7	..
October	102.4	117.6	137.3	158.2	169.9	176.7	183.5	191.4	194.3	..	100.9	111.7	123.8	..
November	103.2	118.8	138.5	160.1	170.9	177.5	184.1	192.1	196.3	..	101.5	112.8	123.4	..
December	104.3	119.8	139.4	161.2	170.5	178.0	183.9	192.4	197.1	..	101.4	113.1	123.3	..

Percentage changes on one year earlier

Tax and price index

	1978	1979	1980	1981	1982	1983	1984	1985	1986	1987	1987	1989	1990	1991
January	4.9	6.1	16.1	14.0	15.6	5.2	4.2	3.8	4.4	2.6		5.6	6.3	8.5
February	4.5	6.5	16.9	13.2	14.4	5.7	4.2	4.3	3.9	..	2.7	6.1	6.2	8.4
March	4.1	6.6	17.6	13.4	13.7	4.8	4.4	5.0	3.0	..	2.8	6.1	6.8	7.8
April	2.1	12.3	18.4	15.7	9.7	3.5	4.1	6.4	1.2	..	2.5	8.3	7.7	6.1
May	1.8	12.6	18.5	15.3	9.8	3.1	4.1	6.5	0.9	..	2.4	8.4	8.1	5.4
June	1.5	13.8	17.4	14.9	9.4	3.0	4.0	6.4	0.6	..	2.5	8.4	8.1	
July	1.9	13.2	18.5	14.3	9.6	3.1	3.3	6.3	0.4	..	2.8	8.5	8.0	
August	2.1	13.4	17.8	14.9	8.7	3.6	3.8	5.5	0.6	..	2.6	7.4	9.0	
September	1.9	14.1	17.3	14.9	7.9	4.2	3.5	5.2	1.2	..	2.4	7.6	9.4	
October	2.0	14.8	16.8	15.2	7.4	4.0	3.8	4.3	1.5	..	2.9	6.0	10.8	
November	4.6	15.1	16.6	15.6	6.7	3.9	3.7	4.3	2.2	..	2.4	6.4	9.4	
December	5.0	14.9	16.4	15.6	5.8	4.4	3.3	4.6	2.4	..	1.9	6.4	9.0	

Retail prices index

	1978	1979	1980	1981	1982	1983	1984	1985	1986	1987	1987	1989	1990	1991
January	9.9	9.3	18.4	13.0	12.0	4.9	5.1	5.0	5.5	3.9		7.5	7.7	9.0
February	9.5	9.6	19.1	12.5	11.0	5.3	5.1	5.4	5.1	..	3.9	7.8	7.5	8.9
March	9.1	9.8	19.8	12.6	10.4	4.6	5.2	6.1	4.2	..	4.0	7.9	8.1	8.2
April	7.9	10.1	21.8	12.0	9.4	4.0	5.2	6.9	3.0	..	4.2	8.0	9.4	6.4
May	7.7	10.3	21.9	11.7	9.5	3.7	5.1	7.0	2.8	..	4.1	8.3	9.7	5.8
June	7.4	11.4	21.0	11.3	9.2	3.7	5.1	7.0	2.5	..	4.2	8.3	9.8	
July	7.8	15.6	16.9	10.9	8.7	4.2	4.5	6.9	2.4	..	4.4	8.2	9.8	
August	8.0	15.8	16.3	11.5	8.0	4.6	5.0	6.2	2.4	..	4.4	7.3	10.6	
September	7.8	16.5	15.9	11.4	7.3	5.1	4.7	5.9	3.0	..	4.2	7.6	10.9	
October	7.8	17.2	15.4	11.7	6.8	5.0	5.0	5.4	3.0	..	4.5	7.3	10.9	
November	8.1	17.4	15.3	12.0	6.3	4.8	4.9	5.5	3.5	..	4.1	7.7	9.7	
December	8.4	17.2	15.1	12.0	5.4	5.3	4.6	5.7	3.7	..	3.7	7.7	9.3	

Note: The purpose and methodology of the Tax and price index were described in an article in the August 1979 issue of *Economic Trends* and in the September *Economic Progress Report* published by the Treasury. The purpose is to produce a single index which measures changes in both direct taxes (including national insurance contributions) and in retail prices for a representative cross-section of taxpayers. Thus, while the Retail prices index may be used to measure changes in the purchasing power of after-tax income (and of the income of non-taxpayers) the Tax and price index takes account of the fact that taxpayers will have more or less to spend according to changes in direct taxation. The index measures the change in gross taxable income which would maintain after tax income in real terms.

The months April, May and June for the years 1979 and 1980 are affected by the late timing of the 1979 Budget.

Source: Central Statistical Office

From: Monthly Digest of Statistics, June 1991, Table 18.5

8.3 Purchasing power of the 1951 pound

United Kingdom

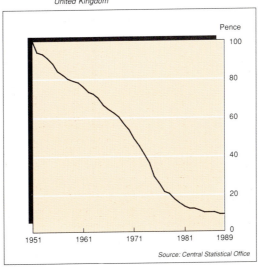

Source: Central Statistical Office

From: Social Trends, 1991, Chart 6.11

PRICES

8.4 Purchasing power of the pound in the European Community[1], 1981 and 1988

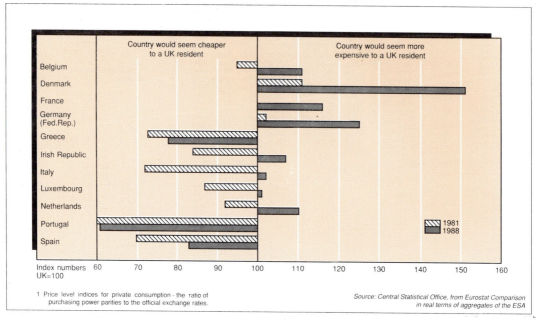

1 Price level indices for private consumption - the ratio of purchasing power parities to the official exchange rates.

Source: Central Statistical Office, from Eurostat Comparison in real terms of aggregates of the ESA

From: Social Trends, 1991, Chart 6.12

8.5 Consumer credit amount outstanding

Great Britain

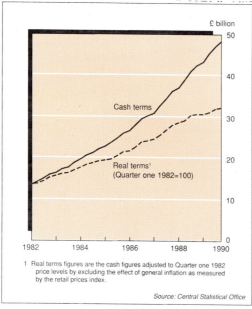

1 Real terms figures are the cash figures adjusted to Quarter one 1982 price levels by excluding the effect of general inflation as measured by the retail prices index.

Source: Central Statistical Office

From: Social Trends, 1991, Chart 6.15

8.6 Personal customers cash withdrawals[1]: by type

Great Britain

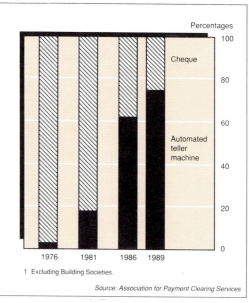

1 Excluding Building Societies.

Source: Association for Payment Clearing Services

From: Social Trends, 1991, Chart 6.17

8.7 Index numbers of producer prices

1985=100, monthly averages

	Materials and fuel purchased[1,2]													
	Manufac-turing industry	Materials	Fuel	Manufac-turing industry (season-ally adjusted)	Manufac-turing other than food, drink and tobacco	Food, drink and tobacco manufac-turing indus-tries	Metal manufac-turing	Extrac-tion of minerals not elsewhere specified	Non-metallic mineral products	Chemical industry	Man-made fibres	Metal goods, engineer-ing and vehicle indus-tries	Metal goods not elsewhere specified	Mechan-ical engin-eering
SIC 1980 Division Class or Group	2 to 4		2 to 4		2 to 4 excl. 41/42	41/42	22	23	24	25	26	3	31	32
	DZBR	DZBS	DZBT	DZDQ	DZBU	DZBX	DZBY	DZBZ	DZCA	DZCB	DZCC	DZCD	DZCE	DZCF
1986	92.4	93.6	87.2	92.3	89.5	98.8	95.3	89.2	95.2	82.9	91.2	98.4	97.9	101.4
1987	95.3	97.4	85.5	95.3	94.0	99.5	99.2	91.4	98.0	86.8	97.8	101.6	101.0	104.4
1988	98.4	101.5	84.3	98.4	98.4	100.9	110.3	91.7	101.2	85.7	100.6	110.0	111.4	111.6
1989	104.0	107.5	88.9	104.0	103.7	107.1	116.3	97.4	105.8	90.2	103.6	116.3	117.5	118.9
1990	103.8	106.6	91.3	103.8	102.8	109.2	111.8	103.0	110.6	95.2	105.4	114.9	114.2	122.5
1987 Sep	95.1	98.1	81.6	96.3	94.6	98.3	101.1	90.7	98.1	86.3	99.6	102.6	102.2	105.2
Oct	95.6	98.7	81.2	96.5	95.1	98.7	101.1	90.6	98.1	86.4	98.5	103.3	103.5	105.7
Nov	95.6	97.5	87.1	95.7	94.7	99.5	100.4	92.1	99.6	85.5	98.9	103.7	103.8	106.0
Dec	97.9	98.2	96.4	95.3	97.2	100.9	103.8	95.1	101.5	86.0	100.2	105.9	106.0	107.1
1988 Jan	98.1	98.5	96.0	95.4	98.1	100.3	105.3	95.4	101.8	86.8	100.8	107.1	107.1	108.4
Feb	96.8	99.4	84.7	95.7	96.6	99.5	104.5	90.7	99.4	85.8	99.7	106.3	107.3	108.5
Mar	95.7	99.3	79.2	95.8	95.4	99.2	104.6	88.3	99.0	83.7	98.7	106.4	107.5	108.6
Apr	96.3	99.6	81.6	96.5	96.1	99.7	104.6	89.8	99.7	84.8	99.3	107.2	107.9	109.5
May	97.7	101.1	82.1	97.9	97.4	100.7	109.4	91.0	100.4	85.6	99.2	108.5	110.0	110.8
Jun	99.5	103.5	81.5	100.1	99.9	101.2	116.6	90.9	100.8	86.2	98.7	110.9	113.2	111.9
Jul	99.4	103.4	81.0	100.1	98.8	102.8	111.4	91.2	100.7	86.1	99.8	110.4	112.1	112.1
Aug	98.8	102.6	81.3	100.1	99.0	101.1	112.3	91.7	101.1	85.8	100.0	110.9	112.8	112.7
Sep	98.2	102.1	80.3	99.3	98.3	100.8	111.2	90.9	101.2	85.4	99.9	110.9	112.8	113.3
Oct	98.0	101.9	80.0	98.9	98.4	100.4	111.8	90.5	101.7	85.0	102.6	111.7	113.8	113.7
Nov	99.8	102.8	86.3	99.8	100.0	102.1	114.1	92.8	103.3	85.8	103.4	113.3	115.1	114.5
Dec	102.6	103.6	97.7	100.4	103.3	103.4	117.4	97.7	105.8	87.6	104.6	115.8	117.4	115.8
1989 Jan	104.0	105.2	98.7	101.3	105.2	103.9	119.9	99.2	106.8	89.8	105.4	117.0	118.6	117.4
Feb	101.9	105.2	87.6	101.0	102.6	103.5	115.9	95.0	104.4	88.6	103.7	115.4	116.7	117.2
Mar	102.4	106.9	82.4	102.6	102.6	104.7	116.6	93.5	105.5	89.3	103.1	115.7	117.5	117.4
Apr	103.9	108.1	84.9	104.0	103.5	107.0	116.8	95.2	106.5	90.5	104.2	116.2	117.8	118.3
May	104.7	109.1	85.2	105.0	104.3	107.8	117.8	96.5	107.3	91.2	104.4	116.7	118.4	119.1
Jun	104.7	109.1	84.8	105.3	103.9	108.2	116.6	96.4	105.9	91.3	104.6	116.7	118.3	119.4
Jul	102.8	106.9	84.8	104.5	101.6	107.7	112.4	95.8	103.4	89.3	103.9	115.2	116.2	118.9
Aug	102.7	106.8	84.6	104.0	102.2	106.5	115.0	95.5	103.2	88.7	102.2	116.0	117.3	119.3
Sep	103.8	108.0	85.2	104.8	103.0	107.8	116.2	96.4	103.9	89.4	102.4	116.5	117.9	119.7
Oct	104.1	108.1	86.5	104.8	102.8	108.8	116.1	97.4	104.4	89.3	101.7	116.1	117.4	119.7
Nov	105.7	108.3	94.4	105.6	104.8	109.6	116.2	101.0	107.3	91.4	103.1	116.4	116.9	120.0
Dec	107.7	107.9	107.2	105.3	107.6	110.1	116.2	106.5	110.6	93.9	105.0	117.3	117.0	120.7
1990 Jan	107.4	107.5	107.8	104.7	107.5	110.1	113.9	108.1	111.8	95.9	106.8	116.7	116.0	121.8
Feb	104.6	106.9	94.9	103.8	103.5	109.8	109.6	102.4	108.8	93.4	104.8	114.2	113.4	120.9
Mar	105.1	108.8	88.8	105.2	103.7	109.8	113.9	100.1	109.4	91.8	104.1	115.6	114.9	121.4
Apr	104.7	108.8	86.2	104.7	102.4	112.0	113.0	99.4	109.4	91.1	104.4	115.4	115.0	122.2
May	103.6	107.8	84.5	103.8	101.3	111.0	113.0	98.8	108.7	89.8	104.0	115.0	114.9	122.2
Jun	102.1	106.3	83.2	102.8	99.8	110.1	111.7	98.0	108.4	88.4	103.6	114.1	113.9	122.1
Jul	101.1	104.9	83.5	102.7	98.8	109.2	110.5	98.0	108.5	88.6	103.0	114.0	113.3	122.3
Aug	101.9	105.3	87.1	102.9	101.6	106.6	111.4	101.7	110.1	96.1	103.4	114.4	113.8	122.5
Sep	104.1	107.6	88.4	104.8	104.7	106.8	114.8	104.0	111.3	101.4	103.9	115.6	115.3	123.4
Oct	103.4	106.6	91.3	104.1	104.1	106.8	110.8	106.6	112.4	103.6	107.5	114.6	113.6	123.4
Nov	103.0	104.7	95.4	102.9	102.7	107.7	108.4	107.8	112.9	101.3	108.9	114.2	112.4	123.4
Dec	104.7	104.7	104.9	102.9	104.1	109.5	110.1	111.3	115.2	100.6	110.2	115.5	113.3	124.2
1991 Jan	104.4	104.6	104.0	102.5	103.4	110.2	110.0	112.2	115.8	100.9	108.4	115.4	112.9	125.5
Feb	102.3	103.5	97.1	101.8	99.9	110.5	107.5	108.8	114.3	95.9	106.2	114.4	112.4	125.3
Mar	102.4	105.3	89.2	102.3	99.4	111.9	109.6	104.5	112.7	95.3	105.0	114.3	112.8	125.6
Apr	103.5	106.5	90.3	103.5	100.6	112.9	112.0	104.9	112.8	95.4	104.0	115.2	113.7	126.5
May	103.5	106.6	89.9	103.7	100.3	113.3	110.8	105.2	113.0	96.3	103.9	114.7	113.5	126.5

Note: The dagger symbol beside a figure indicates the earliest revised value for each series. Figures for the last 2 months shown are provisional.
1 Index numbers are constructed on a net sector basis ie transactions within sector are excluded.

2 Index numbers are compiled exclusive of VAT. Revenue duties (on cigarettes, tobacco and alcoholic liquor) are included, as is duty on hydrocarbon oils.

Source: Central Statistical Office

From: Monthly Digest of Statistics, June 1991, Table 18.6

PRICES

8.7 Index numbers of producer prices (cont.)

Price index numbers of materials and fuel purchased[1,2]

	Electrical and electronic engineering	Motor vehicles and parts	Other transport equipment	Instrument engineering	Food manufacturing industries	Materials	Fuel	Textile industry	Footwear and clothing industries	Timber and wooden furniture industries	Paper and paper products	Processing of rubber and plastics	Other manufacturing industries	Construction materials	House-building materials
SIC 1980 Division Class or Group	34	35	36	37	411 to 423			43	45	46	47	48	49	5	part of 5
	DZCG	DZCH	DZCI	DZCJ	DZCK	DZCL	DZCM	DZCN	DZCO	DZCP	DZCQ	DZCR	DZCS	DZCT	DZCU
1986	99.4	102.3	101.6	101.7	99.1	99.9	84.4	97.4	100.0	99.6	100.1	98.2	98.6	103.6	103.8
1987	104.1	106.8	105.1	105.8	99.5	100.4	83.3	103.2	104.5	105.5	108.1	104.8	104.9	109.3	110.1
1988	113.0	113.0	112.1	112.6	101.0	102.0	81.2	107.9	106.9	109.4	113.5	111.1	109.4	115.4	116.2
1989	119.4	119.5	118.3	119.0	107.5	108.6	85.9	115.0	109.8	116.9	120.1	111.7	114.7	123.4	124.1
1990	120.3	124.2	123.2	123.9	109.2	110.3	89.1	112.8	113.6	126.6	123.2	113.9	115.7	129.6	130.2
1987 Sep	105.3	107.7	105.7	106.9	98.1	99.1	79.9	105.2	105.7	106.3	108.7	106.5	106.9	110.7	111.6
Oct	106.3	108.0	106.2	107.5	98.7	99.7	79.3	105.9	105.9	106.6	109.2	107.0	107.6	111.1	112.0
Nov	106.8	108.4	106.7	107.8	99.5	100.3	84.1	104.9	105.6	106.8	109.8	107.7	106.4	111.6	112.5
Dec	108.9	109.4	107.9	108.6	101.0	101.5	92.2	105.7	105.9	107.6	111.1	109.0	107.9	112.2	113.1
1988 Jan	110.0	110.6	109.1	109.8	100.3	100.7	91.7	106.1	106.4	108.1	113.2	110.0	108.2	112.7	113.6
Feb	109.2	110.6	109.2	110.0	99.5	100.4	81.3	106.0	106.4	107.8	112.5	109.4	106.9	112.5	113.3
Mar	109.5	110.8	110.1	110.4	99.4	100.6	76.3	105.6	106.3	107.8	111.8	108.8	106.6	113.2	114.0
Apr	110.0	111.5	111.2	111.0	100.0	101.1	78.8	106.4	107.0	108.2	112.3	109.9	106.4	114.0	114.8
May	111.9	112.3	111.8	112.1	100.8	101.9	79.6	106.5	106.9	108.5	112.3	111.2	107.9	114.2	114.9
Jun	114.8	113.2	112.5	113.2	101.5	102.7	79.1	106.8	106.8	108.9	112.5	111.6	110.6	114.8	115.5
Jul	113.2	113.0	112.4	112.9	103.0	104.3	78.5	107.7	107.0	109.4	113.4	111.8	110.3	115.5	116.5
Aug	113.6	114.0	112.9	113.3	100.9	102.1	78.9	108.1	107.1	109.8	113.5	111.9	110.0	116.0	116.9
Sep	113.7	114.0	112.8	113.6	100.6	101.8	77.5	109.6	107.0	110.0	113.4	111.3	110.1	116.9	117.7
Oct	114.7	114.3	113.0	114.0	100.3	101.5	77.0	109.9	107.2	110.4	114.7	111.3	110.9	117.9	118.7
Nov	116.5	115.4	114.1	115.1	102.3	103.3	82.5	110.4	107.2	111.2	115.3	112.3	112.3	118.5	119.2
Dec	118.5	116.6	115.8	116.2	103.5	104.0	92.8	111.4	107.3	112.3	116.8	113.4	113.2	118.8	119.6
1989 Jan	119.9	118.2	117.2	117.3	104.2	104.9	94.0	113.8	108.2	113.5	117.9	114.2	114.0	120.1	120.9
Feb	118.5	118.0	117.0	117.2	103.7	104.7	84.1	112.0	108.1	113.2	117.2	112.7	112.7	120.6	121.5
Mar	119.0	118.0	117.1	117.4	105.1	106.4	79.4	112.2	107.7	113.5	117.3	112.0	113.8	122.3	123.0
Apr	119.3	118.6	117.6	118.2	107.6	109.0	82.0	112.4	108.3	114.2	119.2	112.6	114.4	122.9	123.5
May	119.6	119.1	118.1	118.9	108.4	109.8	82.9	113.8	108.8	115.2	119.5	112.3	114.3	123.1	123.8
Jun	119.3	119.2	118.2	119.1	108.7	110.1	82.5	116.0	109.7	116.2	119.9	111.6	114.8	123.1	123.6
Jul	118.0	119.1	118.2	118.9	107.8	109.1	82.5	113.7	110.0	117.0	120.6	110.5	113.4	123.4	124.0
Aug	119.3	119.6	118.6	119.2	106.4	107.7	82.0	114.3	110.2	117.9	120.6	109.9	114.5	124.1	124.8
Sep	119.9	120.4	118.8	119.7	106.0	108.0	82.7	116.8	110.7	118.7	121.0	109.9	115.5	125.1	125.8
Oct	120.0	120.6	119.0	120.1	109.3	110.6	84.2	116.8	111.3	119.4	121.1	110.0	115.5	125.3	126.1
Nov	119.9	121.1	119.8	120.5	110.0	111.0	91.3	118.1	112.0	120.9	122.4	111.4	116.5	125.4	126.1
Dec	120.1	122.0	120.5	121.0	110.4	110.8	102.8	119.6	112.4	122.6	124.3	113.1	116.8	125.3	126.2
1990 Jan	120.0	123.3	121.7	122.5	110.3	110.6	103.6	120.3	113.9	124.4	125.6	114.4	116.5	126.5	127.4
Feb	118.6	122.6	121.1	122.0	110.0	110.4	91.8	118.1	113.6	124.1	123.8	112.9	115.4	126.6	127.6
Mar	120.6	122.9	121.5	122.7	111.3	112.6	86.3	115.7	113.6	124.8	123.5	112.3	116.8	128.2	128.7
Apr	120.8	123.5	122.4	123.3	112.5	114.0	84.2	115.8	114.3	125.7	122.8	112.3	116.9	129.5	130.0
May	120.6	123.5	122.5	123.4	111.5	113.1	82.4	113.8	113.5	126.2	123.0	112.3	116.5	129.8	130.2
Jun	119.9	123.4	122.6	123.4	110.4	111.9	81.1	112.8	113.4	126.6	122.6	111.9	115.3	129.7	130.1
Jul	119.8	123.8	122.9	123.6	109.0	110.6	81.1	110.4	113.2	126.8	122.8	111.9	115.2	130.4	130.9
Aug	120.5	124.3	123.6	124.1	106.1	107.2	85.3	109.3	113.0	127.3	122.7	112.2	115.6	130.7	131.1
Sep	121.7	125.1	124.2	124.8	106.5	107.5	87.4	110.1	113.3	127.7	123.1	113.5	116.3	130.7	131.1
Oct	120.4	125.4	124.9	125.2	106.2	107.1	90.4	109.6	113.3	128.0	122.3	115.5	114.6	131.0	131.5
Nov	119.8	125.6	125.2	125.4	107.3	108.1	93.6	108.2	113.7	128.3	123.1	118.0	114.0	131.1	131.6
Dec	120.5	126.4	126.3	126.2	109.3	109.7	101.9	110.0	114.3	129.0	123.6	119.6	114.8	131.0	131.6
1991 Jan	121.5	127.8	127.8	127.5	109.9	110.4	101.0	110.3	115.1	129.4	123.1	119.6	115.3	132.6	133.0
Feb	120.8	127.5	127.8	127.5	110.6	111.5	94.3	108.1	114.9	129.0	122.6	117.7	114.3	132.6	133.2
Mar	120.8	128.0	127.4	127.5	112.0	113.3	86.9	107.7	114.9	128.5	120.8	115.4	115.4	133.8	134.0
Apr	121.5	128.7	128.2	128.3	113.1	114.4	88.2	108.9	115.2	128.5	121.7	113.3	116.6	134.1	134.3
May	121.0	128.6	128.3	128.2	113.5	114.8	87.7	109.6	115.5	128.6	121.5	112.0	116.1	134.4	134.7

Note: The dagger symbol beside a figure indicates the earliest revised value for each series. Figures for the last 2 months shown are provisional.

1 Index numbers are constructed on a net sector basis ie transactions within sector are excluded.

2 Index numbers are compiled exclusive of VAT. Revenue duties (on cigarettes, tobacco and alcoholic liquor) are included, as is duty on hydrocarbon oils.

Source: Central Statistical Office

From: Monthly Digest of Statistics, June 1991, Table 18.6

Definitions and sources

The household sector includes private trusts and individuals living in institutions as well as those living in households. It differs from the personal sector, as defined in the national accounts, in that it excludes unincorporated private businesses, private non-profit-making bodies serving persons, and funds of life assurance and pension schemes. More information is given in an article in *Economic Trends*, September 1981.

Household disposable income is equal to the total current income of the household sector *less* payment of United Kingdom taxes on income, employees' national insurance contributions, and contributions of employees to occupational pension schemes. It is revalued at constant prices by a deflator implied by estimates of total household expenditure at current and constant prices. This deflator is a modified form of the consumers' expenditure deflator.

Original income is defined as income before government intervention (i.e. before adding social security benefits etc.) Final income is equal to total current income *less* income tax, employees' national insurance contributions and indirect taxes such as VAT, *plus* the imputed benefits received from certain government expenditure programmes such as the health and education services. For more information and definitions see article 'The effects of taxes and benefits on household income', *Economic Trends*, May 1990.

For other sources see:

Guide to Official Statistics, 1990 edition (200 pages approximately fully indexed) HMSO.

Inland Revenue Statistics, HMSO.

Family Expenditure Survey, HMSO.

9.1 Average earnings index: all employees: main industrial sectors
Great Britain   Classified according to the Standard Industrial Classification 1980

	Whole economy (Divisions 0-9)			Manufacturing industries (Divisions 2-4)			Production industries (Divisions 1-4)			Service industries (Divisions 6-9)		
	Actual	Seasonally adjusted	Underlying trend	Actual	Seasonally adjusted	Underlying trend	Actual	Seasonally adjusted	Underlying trend	Actual	Seasonally adjusted	Underlying trend
1985=100												
	DNFV			DNFW			DNFX			DNFY		
1985	100.0			100.0			100.0			100.0		
1986	107.9			107.7			108.0			107.7		
1987	116.3			116.3			116.7			116.0		
1988	126.4			126.2			126.5			126.2		
1988=100												
	DNAA	DNAB	DNEM	DNAC	DNAD	DNEO	DNAE	DNAF	DNEN	DNDU	DNDV	DNDX
1988	100.0	100.0	..	100.0	100.0	..	100.0	100.0	..	100.0	100.0	..
1989	109.1	109.1	..	108.7	108.7	..	109.1	109.1	..	108.9	108.9	..
1990	119.7	119.7	..	118.9	119.0	..	119.4	119.4	..	119.4	119.4	..
1988 May	98.4	98.5	8.50	99.3	99.2	8.75	99.5	99.9	8.50	98.0	98.3	8.50
Jun	99.8	99.2	8.75	100.6	99.3	9.00	100.4	99.2	9.00	99.6	99.8	8.75
Jul	101.3	100.2	9.00	101.1	100.0	9.00	101.3	100.2	9.00	101.3	100.0	9.00
Aug	100.3	100.1	9.25	100.5	100.4	8.75	100.9	100.6	9.00	100.5	99.7	9.25
Sep	100.9	101.1	9.25	100.2	101.2	8.75	100.5	101.4	8.75	100.6	100.5	9.25
Oct	101.7	102.2	9.00	101.8	102.2	8.50	101.9	102.6	8.75	101.2	101.7	9.00
Nov	103.7	103.3	8.75	103.6	103.1	8.75	103.7	103.1	8.75	103.6	103.7	8.75
Dec	106.9	105.8	8.75	105.5	104.6	8.75	105.3	104.6	9.00	107.9	106.3	8.75
1989 Jan	104.2	105.4	9.00	104.2	104.7	8.75	104.2	104.6	8.75	104.2	105.5	9.00
Feb	104.6	106.1	9.25	105.0	105.8	8.50	104.9	105.6	8.75	104.4	105.6	9.25
Mar	107.3	107.3	9.50	105.7	105.6	8.75	106.0	105.8	8.75	107.8	107.8	9.50
Apr	107.3	107.4	9.25	108.2	107.9	8.50	107.9	108.0	8.75	107.1	107.3	9.25
May	107.5	107.6	9.00	108.0	107.9	8.75	108.1	108.5	8.75	107.2	107.5	9.00
Jun	109.1	108.4	8.75	109.4	108.0	8.50	109.6	108.2	8.75	108.5	108.7	8.50
Jul	110.3	109.1	8.75	110.3	109.2	8.50	110.8	109.5	9.00	109.7	108.4	8.25
Aug	109.1	108.9	8.75	108.3	109.3	8.75	109.2	110.0	9.25	108.7	107.8	8.50
Sep	110.7	110.9	9.00	109.5	110.5	8.75	109.8	110.8	9.00	110.4	110.3	8.75
Oct	111.7	112.2	9.25	110.6	111.0	9.00	111.0	111.8	9.25	111.6	112.2	9.00
Nov	113.2	112.8	9.25	112.2	111.6	8.75	112.9	112.2	9.00	112.7	112.7	9.25
Dec	114.7	113.5	8.75	113.8	112.9	8.50	114.3	113.5	9.00	114.3	112.7	9.00
1990 Jan	113.8	115.1	9.50	112.7	113.2	8.75	113.2	113.6	9.25	113.9	115.2	9.25
Feb	114.0	115.6	9.50	113.9	114.7	9.25	114.3	115.0	9.50	113.7	115.0	9.25
Mar	117.4	117.3	9.50	116.8	116.8	9.50	117.0	116.8	9.75	117.2	117.2	9.25
Apr	117.3	117.4	9.75	117.2	117.6	9.50	117.4	117.6	9.75	116.9	117.2	9.50
May	118.5	118.7	9.75	117.8	117.9	9.25	118.2	118.6	9.75	118.6	118.9	9.75
Jun	120.5	119.8	10.00	120.1	118.6	9.50	120.7	119.3	9.75	119.8	120.1	10.00
Jul	121.2	119.9	10.25	120.8	119.6	9.50	121.3	119.9	10.00	120.5	119.1	10.00
Aug	120.9	120.7	10.00	118.8	119.9	9.50	119.7	120.6	9.75	121.1	120.2	10.00
Sep	121.3	121.5	10.00	120.2	121.4	9.50	121.0	122.1	9.75	120.6	120.5	10.00
Oct	121.7	122.3	9.75	120.8	121.2	9.25	121.6	122.4	9.75	120.9	121.5	9.75
Nov	123.8	123.3	9.75	122.4	122.4	9.50	123.7	122.9	9.75	123.0	123.1	9.75
Dec	126.3	125.0	9.75	125.1	124.1	9.50	125.2	124.4	9.75	126.3	124.5	9.50
1991 Jan	124.3	125.7	9.50	123.4	123.9	9.25	124.3	124.7	9.50	123.8	125.3	9.50
Feb	124.7	126.4	9.25	124.3	125.2	8.75	125.2	126.0	9.00	123.8	125.2	9.00
Mar	127.5	127.5	9.00	126.1	126.0	8.50	126.8	126.6	9.00	127.6	127.6	8.75
Apr	127.2	127.3	8.75	128.0	128.4	8.50	128.5	128.7	9.00	125.9	126.2	8.50

The seasonal adjustment factors currently used for the SIC 1980 series are based on data up to January 1988.

See note on page 121.

Source: Department of Employment

From: *Monthly Digest of Statistics, June 1991 Table 18.11*

INCOMES

9.2 Gross weekly earnings of full-time employees [1]: by sex and type of employment
Great Britain   £s and percentages

	Males					Females				
	1971	1981	1986	1988	1989	1971	1981	1986	1988	1989
Manual employees										
Mean (£)	29.0	120.2	174.4	200.6	217.8	15.3	74.7	107.5	123.6	134.9
Median (£)	27.7	112.8	163.4	188.0	203.9	14.6	71.6	101.1	115.6	125.9
As percentage of median										
Highest decile	*147*	*151*	*155*	*157*	*158*	*143*	*143*	*150*	*154*	*156*
Lowest decile	*68*	*69*	*65*	*64*	*63*	*71*	*70*	*69*	*69*	*69*
Non-manual employees										
Mean (£)	38.5	160.5	244.9	294.1	323.6	20.0	97.5	145.7	175.5	195.0
Median (£)	34.0	147.0	219.4	259.7	285.7	18.2	87.7	131.5	157.1	173.5
As percentage of median										
Highest decile	*175*	*167*	*175*	*180*	*181*	*169*	*172*	*167*	*172*	*174*
Lowest decile	*60*	*60*	*57*	*55*	*54*	*65*	*68*	*65*	*63*	*62*
All employees										
Mean (£)	32.4	138.2	207.5	245.8	269.5	18.4	92.0	137.2	164.2	182.3
Median (£)	29.4	124.6	185.1	215.5	235.5	16.7	82.8	123.4	145.3	160.1
As percentage of median										
Highest decile	*162*	*168*	*173*	*178*	*180*	*165*	*172*	*170*	*178*	*181*
Lowest decile	*65*	*64*	*60*	*59*	*59*	*66*	*68*	*65*	*63*	*63*

1 Figures relate to April each year and to full-time employees on adult
  rates whose pay for the survey pay period was not affected by absence.

Source: New Earnings Survey, Employment Department

From: Social Trends, 1991, Table 5.5

9.3 Net income after rent and community charge [1]: by level of gross earnings and type of family, 1990[2]
Great Britain

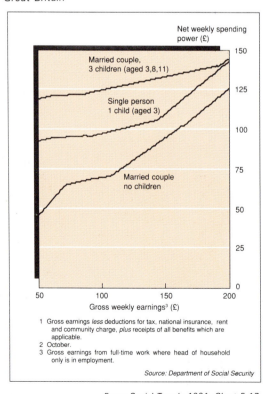

1 Gross earnings *less* deductions for tax, national insurance, rent
  and community charge, *plus* receipts of all benefits which are
  applicable.
2 October.
3 Gross earnings from full-time work where head of household
  only is in employment.

Source: Department of Social Security

From: Social Trends 1991, Chart 5.17

9.4 Percentage of income paid in income tax and national insurance contributions [1]: by marital status and level of earnings [2]

United Kingdom · Percentages

	1981–82	1983–84	1984–85	1985–86	1986–87	1987–88	1988–89	1989–90[3]	1990–91[4]
Single person									
Half average earnings									
Tax	17.5	16.0	15.3	14.9	14.5	13.9	13.0	13.2	13.4
NIC	7.7	9.0	9.0	7.0	7.0	7.0	7.0	6.7	6.4
Average earnings									
Tax	23.7	23.0	22.7	22.4	21.7	20.4	19.0	19.1	19.2
NIC	7.7	9.0	9.0	9.0	9.0	9.0	9.0	8.3	7.7
Twice average earnings									
Tax	27.3	26.5	26.3	26.2	25.4	23.7	22.0	22.1	23.3
NIC	6.1	7.2	7.1	7.1	7.2	6.9	6.6	6.1	5.7
Married man[5]									
Half average earnings									
Tax	10.5	8.0	6.9	6.3	6.3	6.5	6.1	6.4	6.7
NIC	7.7	9.0	9.0	7.0	7.0	7.0	7.0	6.7	6.4
Average earnings									
Tax	20.2	19.0	18.5	18.1	17.6	16.7	15.5	15.7	15.9
NIC	7.7	9.0	9.0	9.0	9.0	9.0	9.0	8.3	7.7
Twice average earnings									
Tax	25.1	24.5	24.2	24.1	23.3	21.9	20.3	20.4	20.4
NIC	6.1	7.2	7.1	7.1	7.2	6.9	6.6	6.1	5.7

1 Employees' contributions. Assumes contributions at Class 1, contracted in, standard rate.
2 Earnings are average earnings for full-time adult male manual employees working a full week on adult rates.
3 Average for financial year.
4 1989-90 based projections
5 Assuming wife not in paid employment.

Source: Inland Revenue

From: Social Trends 1991, Table 5.13

9.5 Composition of the lowest quintile group of household income: by economic status of family head[2], 1981 and 1987

Great Britain

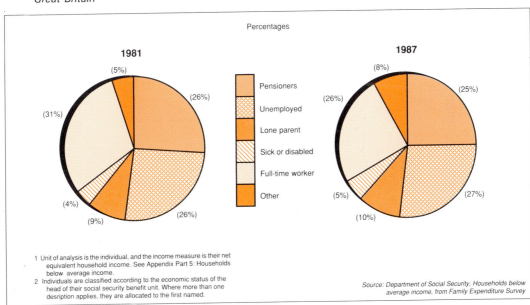

Percentages

1981   1987

(5%)   (8%)
(26%)   (25%)
(31%)   (26%)

Pensioners
Unemployed
Lone parent
Sick or disabled
Full-time worker
Other

(4%)   (5%)
(9%)   (10%)
(26%)   (27%)

1 Unit of analysis is the individual, and the income measure is their net equivalent household income. See Appendix Part 5: Households below average income.
2 Individuals are classified according to the economic status of the head of their social security benefit unit. Where more than one desription applies, they are allocated to the first named.

Source: Department of Social Security, Households below average income, from Family Expenditure Survey

From: Social Trends, 1991, Chart 5.18

INCOMES

9.6 Household income [1]

United Kingdom

	1971	1976	1981	1983	1984	1985	1986	1987	1988	1989
Source of income *(percentages)*										
Wages and salaries[2]	68	67	63	61	61	60	59	59	59	59
Income from self-employment[3]	9	9	8	9	9	9	10	10	11	11
Rent, dividends, interest	6	6	7	6	6	7	7	7	7	8
Private pensions, annuities, etc	5	5	6	7	7	8	8	8	8	8
Social security benefits	10	11	13	14	13	13	13	13	12	11
Other current transfers[4]	2	2	3	3	4	3	3	3	3	3
Total household income (= 100%) (£ billion)	44.7	100.4	202.4	238.8	258.1	283.5	311.5	337.3	374.4	420.0
Direct taxes etc *(percentages of total household income)*										
Taxes on income	14	17	14	15	15	15	14	14	14	14
National insurance contributions[5]	3	3	3	4	4	4	4	4	4	4
Contributions to pension schemes	1	2	2	2	2	2	2	2	2	2
Total household disposable income (£ billion)	36.4	78.3	162.7	189.5	204.8	225.2	249.1	270.3	300.0	337.8
Real household disposable income per head (index numbers — 1985 = 100)	73	81	92	94	96	100	106	110	116	123
Annual change on previous year (percentages)	0.3	− 1.0	− 1.1	1.7	2.6	4.1	5.7	3.7	5.6	6.1

1 See Appendix, Part 5: The household sector.
2 Includes Forces' pay and income in kind.
3 After deducting interest payments, depreciation, and stock appreciation.

4 Mostly other government grants, but including transfers from abroad and from non-profit-making bodies.
5 By employees and the self-employed.

Source: Central Statistical Office

From: Social Trends, 1991, Table 5.2

9.7 Real household disposable income per head
United Kingdom

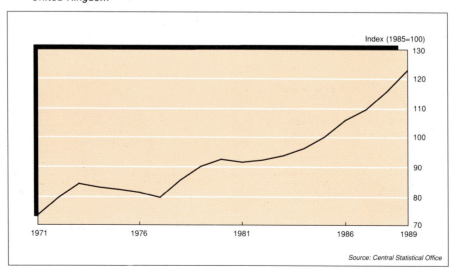

Index (1985=100)

Source: Central Statistical Office

From: Social Trends, 1991, Chart 5.1

9.8 Income of pensioners[1]: by source

United Kingdom

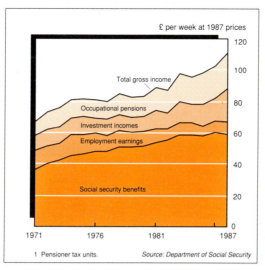

£ per week at 1987 prices

Total gross income

Occupational pensions

Investment incomes

Employment earnings

Social security benefits

1971   1976   1981   1987

1 Pensioner tax units.   *Source: Department of Social Security*

From: Social Trends 1991, Chart 5.6

9.9 Social security benefits[1] for the unemployed[2]: by sex

Great Britain   Percentages and thousands

	1961	1971	1976	1981	1986	1988	1989
Unemployed male claimants receiving each benefit							
(percentages)							
Unemployment benefit only	47.2	40.9	30.0	28.2	20.2	16.3	11.6
Unemployment and supplementary benefit/income support	9.4	13.6	10.9	11.4	7.5	7.2	6.4
Supplementary benefit/income support only	21.9	27.1	42.4	46.0	59.3	61.2	67.5
No benefit	21.6	18.4	16.6	14.4	13.0	15.2	14.5
Total unemployed male claimants (= 100%) (thousands)	283	721	1,076	1,994	2,086	1,344	1,062
Unemployed female claimants receiving each benefit							
(percentages)							
Unemployment benefit only	39.7	41.0	29.1	38.9	33.5	29.7	22.3
Unemployment and supplementary benefit/income support	2.5	6.7	3.9	3.8	2.7	3.1	2.8
Supplementary benefit/income support only	12.2	20.5	44.6	37.4	40.8	43.4	51.7
No benefit	45.5	31.8	22.4	20.0	23.0	23.7	23.2
Total unemployed female claimants (= 100%) (thousands)	101	138	380	709	955	559	391

1 At November each year except for 1976 when figures relate to August and 1981 when figures for 1982 are used. From April 1988 supplementary benefit was replaced by income support.

2 Prior to 1981 count of registered unemployed; for 1981 count of registered unemployed claimants, after then, count of unemployed claimants.

Source: Department of Social Security

From: Social Trends, 1991, Table 5.10

Definition and sources

More information is available in the *Statement on the Defence Estimates 1990* Volume 1 (Cmnd 1022-I) and 2 (Cmnd 1022-II) HMSO.

For other sources see:

Guide to Official Statistics, 1990 Edition (200 pages approximately fully indexed) HMSO.

10.1 Formation of the armed forces *At 1 April*

Number

	Unit[1]	1980	1981	1982	1983	1984	1985	1986	1987	1988	1989	1990
Royal Navy[2]												
Submarines	Vessels	23	22	22	21	22	24	26	26	24	26	24
Carriers and assault ships	Vessels	4	3	3	4	4	4	4	3	3	4	4
Cruisers and destroyers	Vessels	10	12	11	13	13	} 46	47	46	42	39	44
Frigates	Vessels	39	36	38	42	39						
Mine counter-measure[3]	Vessels	35	32	33	36	34	32	39	36	35	36	37
Patrol ships and craft	Vessels	23	22	21	25	31	28	32	28	31	34	32
Fixed wing aircraft	Squadrons	2	3	3	3	3	3	3	3	3	3	3
Helicopters	Squadrons	14	14	15	13	12	14	14	14	14	14	13
Royal Marines	Commandos	4	3	3	3	3	3	3	3	3	3	3
Army[4, 5]												
Royal Armoured Corps	Regiments	19	19	19	19	19	19	19	19	19	19	19
Royal Artillery	Regiments	22	22	22	22	22	22	22	22	22	22	22
Royal Engineers[6]	Regiments	10	11	11	12	12	13	13	13	13	13	13
Infantry[6]	Battalions	56	56	57	56	56	56	56	55	55	55	55
Special Air Service	Regiments	1	1	1	1	1	1	1	1	1	1	1
Army Air Corps	Regiments	6	6	6	4	4	4	4	4	4	4	4
Royal Air Force[4]												
Strike/attack	Squadrons	15	15	12	10	11	11	11	11	11	11	11
Offensive support	Squadrons	5	5	5	5	5	5	5	5	5	5	5
Air defence	Squadrons	9	9	9	9	8	9	9	9	9	9	9
Maritime patrol	Squadrons	4	4	4	4	4	4	4	4	4	4	4
Reconnaissance	Squadrons	5	5	3	2	3	3	3	3	3	3	2
Airborne early warning	Squadrons	1	1	1	1	1	1	1	1	1	1	1
Transport[7]	Squadrons	10	9	10	11	10	11	12	12	12	12	12
Tankers	Squadrons	2	2	2	3	3	3	4	3	3	3	3
Search and rescue	Squadrons	3	3	3	3	2	2	2	2	2	2	2
Surface to air missiles	Squadrons	8	8	8	8	8	8	8	8	8	8	7
Ground defence	Squadrons	6	6	6	6	5	5	5	5	5	6	6

1 The number of personnel and the amount of equipment in each vessel, regiment, etc, varies according to the role currently assigned.
2 Excludes vessels undergoing major refit, conversion, or on stand-by, etc.
3 In 1981 four ex-inshore minesweepers used for training are excluded.
4 Front-line Squadrons.
5 Combat Arm major units only.
6 Includes Gurkhas.
7 Includes helicopters.

Source Ministry of Defence
From: Annual Abstract of Statistics, 1991, Table 7.1

10.2 Defence expenditure [1,2]

£ million

	1978/79	1979/80	1980/81	1981/82	1982/83	1983/84	1984/85	1985/86	1986/87	1987/88	1988/89
Total expenditure at outturn prices[3]	7 455	9 178	11 182	12 607	14 412	15 487	17 122	17 943	18 163	18 856	19 072
of which:											
Expenditure on personnel	3 293	3 912	4 556	5 058	5 455	5 726	5 983	6 379	6 890	7 212	7 572
of the Armed Forces	1 639	2 099	2 460	2 728	2 914	3 076	3 236	3 510	3 787	4 032	4 300
of the retired Armed Forces	432	459	503	624	680	777	828	899	980	1 079	1 060
of civilian staff	1 222	1 354	1 593	1 706	1 861	1 873	1 919	1 970	2 123	2 101	2 212
Expenditure on equipment	2 984	3 640	4 885	5 638	6 297	6 939	7 838	8 193	7 885	8 270	8 038
Sea systems[4]	878	1 110	1 513	1 624	1 730	1 849	2 228	2 499	2 494	2 797	2 633
Land systems	601	740	904	1 101	1 353	1 475	1 638	1 887	1 759	1 700	1 554
Air systems	1 214	1 427	2 059	2 458	2 640	3 057	3 474	3 296	3 090	3 230	3 085
Other	291	363	410	456	574	558	498	511	542	543	766
Other expenditure	1 178	1 625	1 741	1 910	2 659	2 822	3 302	3 370	3 387	3 374	3 462
Works, buildings and land[5]	405	599	623	664	832	1 067	1 271	1 413	1 498	1 453	1 411
Miscellaneous stores and services	773	1 026	1 118	1 246	1 827	1 754	2 031	1 958	1 889	1 921	2 051
Total expenditure at constant (1988/89) prices[3]	17 412	17 941	18 551	18 815	19 960	20 236	21 035	21 005	20 278	19 940	19 072

1 Expenditure as given in the annual Appropriation Accounts for the Defence Votes, Class I.
2 Including expenditure of government departments other than the Ministry of Defence in years when these contributed to the Defence budget.
3 Because of changes in the responsibilities of the Ministry of Defence, expenditures in successive years are not necessarily comparable.
4 The contractorisation of the dockyards in April 1987 has the effect of increasing sea systems expenditure and decreasing expenditure on personnel and miscellaneous stores and services.
5 Including pay of civilian staff employed in works and building.

Source Ministry of Defence
From: Annual Abstract of Statistics, 1991, Table 7.2

10.3 Defence manpower strengths
At 1 April [1]

Thousands

	1979	1980	1981	1982	1983	1984	1985	1986	1987	1988	1989	1990
UK service personnel												
All services:												
Male	299.7	304.4	316.8	311.9	305.2	309.7	309.8	306.5	303.7	300.9	295.4	288.5
Female	15.3	16.2	16.9	15.7	15.4	16.2	16.4	16.0	16.2	15.9	16.3	17.2
Total	315.0	320.6	333.8	327.6	320.6	325.9	326.2	322.5	319.8	316.9	311.6	305.7
Royal Navy:												
Male	61.2	60.5	62.3	61.1	60.1	59.8	59.1	56.8	55.3	54.3	53.5	52.0
Female	3.8	3.8	4.1	4.0	3.9	3.9	3.7	3.4	3.4	3.3	3.5	3.6
Total	65.1	64.4	66.4	65.1	64.0	63.7	62.8	60.3	58.7	57.7	57.0	55.7
Royal Marines:												
Male	7.4	7.6	7.9	7.9	7.8	7.6	7.6	7.6	7.8	7.8	7.7	7.5
Total	7.4	7.6	7.9	7.9	7.8	7.6	7.6	7.6	7.8	7.8	7.7	7.5
Army:												
Male	150.4	152.8	159.4	157.2	152.9	155.0	155.6	154.8	153.2	151.8	149.1	145.9
Female	5.8	6.3	6.6	6.0	6.1	6.6	6.8	6.6	6.5	6.3	6.5	6.9
Total	156.2	159.0	166.0	163.2	159.1	161.5	162.4	161.4	159.7	158.1	155.6	152.8
Royal Air Force:												
Male	80.7	83.5	87.2	85.7	84.5	87.3	87.5	87.2	87.3	87.0	85.1	83.0
Female	5.6	6.1	6.3	5.8	5.4	5.7	6.0	6.0	6.3	6.3	6.3	6.7
Total	86.3	89.6	93.5	91.5	89.8	93.1	93.4	93.2	93.6	93.3	91.4	89.7
Personnel locally entered overseas:												
Total	8.4	8.2	9.7	10.1	10.1	10.1	10.2	10.1	9.8	9.4	9.1	8.9
Regular Reserves: [1]												
Royal Navy	28.4	27.0	26.8	24.8	23.9	23.5	23.3	24.2	24.5	24.7	25.1	25.8
Royal Marines	2.4	2.2	2.2	2.2	2.2	2.2	2.2	2.3	2.3	2.4	2.5	2.6
Army	127.0	133.1	137.5	140.2	138.3	143.2	150.2	153.9	160.4	167.7	175.3	183.4
Royal Air Force	30.8	30.3	30.1	29.3	28.9	29.0	29.8	31.5	33.7	35.4	37.5	40.1
Total	188.5	192.6	196.5	196.4	193.4	198.0	205.5	211.6	220.9	230.2	240.4	252.0
Volunteer Reserves and Auxiliary Forces:												
Royal Navy	5.4	5.1	5.4	5.4	5.4	5.2	5.2	5.5	5.7	5.6	5.6	5.9
Royal Marines	0.8	0.8	0.9	1.0	1.1	1.1	1.1	1.2	1.2	1.3	1.2	1.1
Territorial Army	59.4	63.3	69.5	72.1	72.8	71.4	73.7	77.7	78.5	74.7	72.5	72.5
Ulster Defence Regiment	7.6	7.4	7.5	7.1	7.1	6.8	6.4	6.6	6.5	6.4	6.3	6.2
Home Service Force	–	–	–	–	0.3	0.3	0.9	3.0	3.3	3.1	3.0	3.2
Royal Air Force	0.3	0.5	0.6	0.6	0.8	1.1	1.2	1.4	1.6	1.8	1.7	1.7
Total [1]	73.4	77.1	84.0	86.3	87.5	85.9	88.6	95.4	96.8	92.8	90.4	90.6
Cadet Forces: [1, 2]												
Royal Navy	23.9	29.1	31.1	28.7	29.4	28.4	28.4	28.0	27.5	27.1	26.5	26.2
Army	72.6	74.6	75.1	74.1	74.5	73.8	72.1	70.9	71.2	69.3	65.9	65.7
Royal Air Force	43.5	44.1	44.4	45.9	45.1	44.3	44.1	44.5	47.0	48.3	46.7	44.2
Total	140.0	147.8	150.6	148.7	149.0	146.5	144.7	143.4	145.7	144.6	139.2	136.0

1 A few of the figures for reserves and cadets were collected at irregular intervals and do not necessarily refer to 1 April; they are the latest figures available at that date.

2 Combined Cadet Force cadets are included under the relevant service.

Source Ministry of Defence

From: Annual Abstract of Statistics, 1991, Table 7.3

10.4 Recruitment of UK Service personnel to each Service

Number

	1979/80	1980/81	1981/82	1982/83	1983/84	1984/85	1985/86	1986/87	1987/88	1988/89	1989/90
All services:											
Male	46 206	46 693	21 188	19 342	33 760	32 076	30 407	31 147	31 215	30 862	32 337
Female	4 446	3 795	1 419	2 305	3 231	2 645	2 244	2 902	2 611	3 001	4 075
Total	50 652	50 488	22 607	21 647	36 991	34 721	32 651	34 049	33 826	33 863	36 412
Royal Navy:											
Male	7 701	8 130	3 353	3 078	4 223	4 231	3 987	4 791	4 601	4 598	4 696
Female	825	958	452	506	562	351	289	545	580	700	860
Total	8 526	9 088	3 805	3 584	4 785	4 582	4 276	5 336	5 181	5 298	5 556
Royal Marines:											
Male	1 676	1 674	699	447	447	954	1 093	1 233	991	937	1 063
Total	1 676	1 674	699	447	447	954	1 093	1 233	991	937	1 063
Army:											
Male	27 164	27 241	13 603	11 679	20 811	20 914	19 173	18 718	19 895	19 921	20 398
Female	2 025	1 630	601	1 392	1 537	1 364	1 095	1 200	1 146	1 427	1 686
Total	29 189	28 871	14 204	13 071	22 348	22 278	20 268	19 918	21 041	21 348	22 084
Royal Air Force:											
Male	9 665	9 648	3 533	4 138	8 279	5 977	6 154	6 405	5 728	5 406	6 180
Female	1 596	1 207	366	407	1 132	930	860	1 157	885	874	1 529
Total	11 261	10 855	3 899	4 545	9 411	6 907	7 014	7 562	6 613	6 280	7 709

Source Ministry of Defence

From: Annual Abstract of Statistics, 1991, Table 7.4

Definitions and sources

Differences in the legal and judicial systems of England and Wales, Scotland and Northern Ireland make it impossible to aggregate statistics for the United Kingdom as a whole.

In England and Wales, indictable offences are those which must or may be tried by a jury in a Crown Court, although most are actually tried at a magistrates' court. Notifiable offences recorded by the police have a slightly wider definition than indictable offences in that they also include all criminal damage however small. Standard list offences include all indictable offences *plus* some summary offences but exclude most summary motoring offences and other less serious summary offences such as drunkenness.

For other sources see:

Guide to Official Statistics, 1990 edition (200 pages approximately fully indexed) HMSO.

Criminal Statistics: England and Wales HMSO.

11.1 Notifiable offences recorded by the police
England & Wales

Thousands

	Violence against the person	Sexual offences	Burglary	Robbery	Theft and handling stolen goods	Fraud and forgery	Criminal damage	Other	Total
	BEAB	BEAC	BEAD	BEAE	BEAF	BEAG	BEAH	BEAI	BEAA
1985	121.7	21.5	866.7	27.5	1 884.1	134.8	539.0	16.7	3 611.9
1986	125.5	22.7	931.6	30.0	2 003.9	133.4	583.6	16.7	3 847.4
1987	141.0	25.2	900.1	32.6	2 052.0	133.0	589.0	19.3	3 892.2
1988	158.2	26.5	817.8	31.4	1 931.3	133.9	593.9	22.7	3 715.8
1989	177.0	29.7	825.9	33.2	2 012.8	134.5	630.1	27.6	3 870.7
1990	184.7	29.0	1 006.8	36.2	2 374.4	147.9	733.4	31.1	4 543.6
1986 Q4	32.8	5.8	243.5	8.1	519.2	32.0	157.2	4.5	1 003.1
1987 Q1	29.7	5.1	243.3	8.1	487.3	31.9	141.8	3.8	951.0
Q2	34.7	6.3	222.1	7.7	522.5	32.7	154.7	4.5	985.3
Q3	38.6	7.2	205.3	8.2	513.5	33.5	142.5	5.2	954.0
Q4	38.0	6.6	229.5	8.6	528.7	34.9	150.0	5.7	1 001.9
1988 Q1	35.5	6.3	229.6	7.9	499.5	35.4	151.4	5.5	971.2
Q2	39.1	6.7	198.0	7.5	484.6	34.1	150.1	5.6	925.7
Q3	40.9	7.2	184.0	8.0	468.1	32.5	141.1	5.5	887.4
Q4	42.7	6.3	206.1	8.1	479.1	31.8	151.4	6.1	931.6
1989 Q1	39.3	7.0	213.3	8.2	479.6	32.1	156.8	6.0	942.2
Q2	45.2	7.5	192.7	7.8	499.1	32.8	156.0	6.6	947.8
Q3	48.3	8.0	192.3	8.0	505.2	35.4	152.5	7.3	957.0
Q4	44.2	7.2	227.7	9.1	528.8	34.1	164.8	7.8	1 023.8
1990 Q1	41.0	6.6	252.3	8.4	555.2	34.7	175.4	7.2	1 080.9
Q2	47.0	7.4	231.6	8.4	586.3	34.4	188.6	7.8	1 111.3
Q3	49.5	7.9	233.4	8.9	590.5	37.3	175.0	7.7	1 110.3
Q4	47.2	7.2	289.5	10.5	642.4	41.4	194.4	8.4	1 241.0
1991 Q1	41.7	6.7	298.7	9.6	649.8	41.3	195.6	8.2	1 251.5

Source: Home Office

From: Monthly Digest of Statistics, June 1991, Table 5.1

11.2 Crimes and offences recorded by the police

Scotland   Thousands

	Non-sexual crimes of violence	Crimes of indecency	Crimes of dishonesty	Fire raising, vandalism etc	Other crimes	Motor vehicle offences	Miscellaneous offences	Total crimes and offences (monthly)	Total crimes and offences (annual)
	BEBC	BEBD	BEBE	BEBF	BEBG	BEBI	BEBH	BEBB	BEBA
1983	13.3	5.5	345.2	72.9	14.1	236.6	112.8	800.3	799.6
1984	13.9	5.7	360.6	79.3	17.1	219.6	115.2	811.4	809.5
1985	15.3	5.8	344.0	79.8	19.3	230.3	119.2	813.6	800.4
1986	15.7	5.4	342.5	78.9	21.4	238.1	120.4	823.5	822.4
1987	18.5	5.2	356.7	76.6	24.4	249.6	127.2	858.3	858.2
1988	18.0	5.1	344.7	73.5	28.6	248.6	124.9	843.5	855.6
1989	18.5	5.7	354.2	78.6	34.0	277.8	124.8	893.6	902.0
1990	18.6	6.0	386.2	86.2	39.6	294.1	127.2	957.9	959.1
1986 Q3	4.1	1.4	85.2	18.5	5.4	57.1	30.9	202.5	..
Q4	4.3	1.4	93.5	20.3	6.0	61.8	31.2	218.5	..
1987 Q1	4.2	1.2	84.1	18.2	5.2	60.6	28.7	202.2	..
Q2	4.6	1.5	90.1	20.1	5.8	63.5	32.8	218.5	..
Q3	4.8	1.4	90.6	19.2	6.8	61.6	33.2	217.6	..
Q4	4.9	1.2	91.8	19.1	6.6	63.9	32.5	220.0	..
1988 Q1	4.4	1.3	85.3	18.4	6.0	64.0	30.2	209.6	..
Q2	4.4	1.3	83.6	18.2	6.9	61.3	31.3	207.1	..
Q3	4.5	1.3	85.3	17.6	7.8	58.3	31.6	206.4	..
Q4	4.6	1.2	90.4	19.3	7.9	65.0	31.9	220.3	..
1989 Q1	4.4	1.2	84.9	19.0	7.1	69.7	27.7	214.0	..
Q2	4.9	1.4	88.1	20.0	8.6	71.2	33.0	227.2	..
Q3	4.7	1.6	89.4	19.2	8.3	64.6	32.4	220.2	..
Q4	4.5	1.5	91.8	20.5	10.0	72.2	31.7	232.1	..
1990 Q1	4.5	1.5	94.6	21.6	9.1	72.5	30.3	234.0	..
Q2	4.5	1.7	92.5	21.4	9.8	75.0	32.8	237.7	..
Q3	4.8	1.5	96.2	20.7	10.0	71.6	32.6	237.5	..
Q4	4.8	1.4	102.9	22.5	10.7	75.0	31.5	248.7	..

Components may not add to totals due to separate rounding.

Source: The Scottish Office Home and Health Department

From: Monthly Digest of Statistics, June 1991, Table 5.2

11.3 Offenders sentenced for indictable offences: by type of offence and type of sentence, 1989

England and Wales   Percentages and thousands

	Discharge	Probation/ Supervision	Fine	Community service order	Fully suspended sentence	Immediate custody under 5 years	Immediate custody 5 years and over	Other	Total sentenced (= 100%) (thousands)
Offences									
Violence against the person	17	7	37	7	9	15	1	7	55.7
Sexual offences	9	14	33	1	8	25	8	2	7.3
Burglary	10	14	14	13	7	31	–	5	43.3
Robbery	4	10	2	6	2	55	14	7	4.6
Theft and handling stolen goods	20	13	42	7	6	10	–	3	134.3
Fraud and forgery	18	12	35	8	12	13	–	2	22.4
Criminal damage	19	17	29	7	5	12	–	11	9.4
Drug offences	8	5	62	3	5	13	3	–	22.6
Motoring	3	3	77	4	3	9	–	1	11.3
Other	11	3	57	5	6	14	–	3	28.1

Source: Home Office

From Social Trends 1991, Table 12.15

11.4 Offences of burglary recorded by the police: by offence
England & Wales

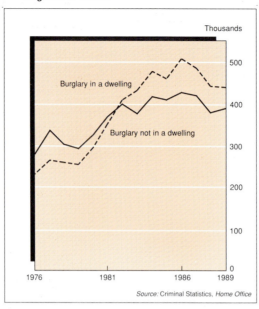

Thousands

Burglary in a dwelling

Burglary not in a dwelling

Source: Criminal Statistics, Home Office

From: Social Trends, 1991, Chart 12.7

11.5 Seizures of controlled drugs and persons found guilty of drug offences, 1989
United Kingdom

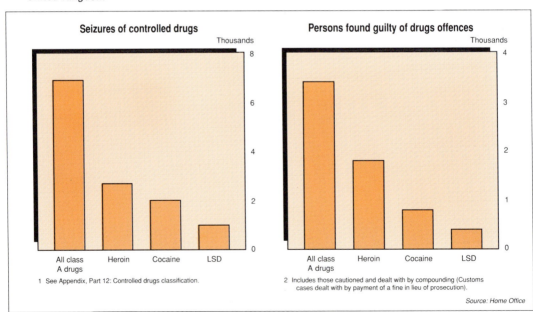

Seizures of controlled drugs

Persons found guilty of drugs offences

Thousands

All class A drugs Heroin Cocaine LSD

1 See Appendix, Part 12: Controlled drugs classification.

2 Includes those cautioned and dealt with by compounding (Customs cases dealt with by payment of a fine in lieu of prosecution).

Source: Home Office

From: Social Trends, 1991, Chart 12.19

11.6 Receptions into prison establishments and population in custody

England & Wales and Northern Ireland
Thousands

	1981	1986	1987	1988	1989
England & Wales and Northern Ireland					
Average population					
Males	44.5	47.0	48.7	49.0	48.5
Females	1.4	1.6	1.7	1.8	1.8
Remand prisoners	7.4	10.2	10.9	10.7	11.0
Untried prisoners	5.3	8.8	9.3	9.0	8.6
Convicted prisoners awaiting sentence[2]	2.1	1.4	1.6	1.7	1.8
Sentenced prisoners[3]	38.1	38.1	39.2	39.8	39.2
Adults	26.7	28.4	29.9	31.2	31.7
Young offenders[4]	11.3	9.7	9.2	8.7	7.5
Other sentences	—	—	—	—	—
Non-criminal prisoners	0.4	0.2	0.3	0.2	0.2
Receptions[1]					
Untried prisoners	49.6	57.6	61.3	59.7	61.6
Convicted prisoners awaiting sentence[2]	24.0	16.6	18.4	17.3	17.6
Sentenced prisoners	90.5	89.8	90.2	85.3	79.9
Non-criminal prisoners	4.8	3.7	3.4	2.9	2.8
England & Wales					
Highest number of inmates sleeping:					
Two in a cell	11.3	13.5	14.0	13.4	12.8
Three in a cell	5.6	4.9	5.3	5.7	5.0

1 Figures of receptions contain an element of double counting as individuals can be received in more than one category of prisoner.
2 Includes persons remanded in custody while social and medical inquiry reports are prepared prior to sentence. Prisoners in Northern Ireland are not committed for sentence but are sentenced at the court of conviction.
3 See Appendix, Part 12: Young offenders.
4 Northern Ireland figures for young prisoners are not available separately before 1980, they are included with adults.

Source: Prison Statistics, *Home Office;*
Northern Ireland Office

From Social Trends 1991, Table 12.22

11.7 Receptions into prison establishments and population in custody

Scotland
Numbers

	1981	1986	1987	1988	1989
Average population	4,518	5,588	5,446	5,229	4,986
Males	4,383	5,394	5,259	5,057	4,838
Females	135	194	187	172	147
Remand prisoners	746	1,017	938	844	770
Untried prisoners	564	796	754	690	639
Convicted prisoners awaiting sentence	182	221	183	153	131
Sentenced prisoners	3,769	4,570	4,509	4,385	4,216
Adults	2,556	3,448	3,474	3,434	3,341
Young offenders[1]	1,174	1,074	987	902	813
Other sentences	41	49	48	48	61
Non-criminal prisoners	1	1	1	1	1
Receptions[2]					
Untried prisoners	10,663	15,295	14,367	13,310	12,177
Convicted prisoners awaiting sentence	2,887	2,812	2,744	1,690	2,104
Sentenced prisoners	15,539	23,220	22,186	20,540	19,484
Non-criminal prisoners	14	13	18	7	15

1 Includes Detention Centre receptions prior to 1989 and Borstal receptions prior to 1984.
2 Total receptions cannot be calculated by adding together receptions in each category because there is double counting. Thus, when a person is received on remand and then under sentence in relation to the same set of charges, he is counted in both categories.

Source: Scottish Home and Health Department

From: Social Trends, 1991 Table 12.23

LAW

11.8 Average population in custody[1]: by type of prisoner
England and Wales

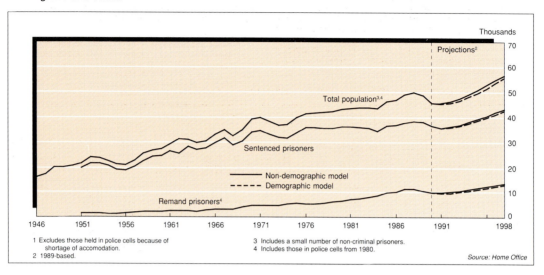

1 Excludes those held in police cells because of shortage of accomodation.
2 1989-based.
3 Includes a small number of non-criminal prisoners.
4 Includes those in police cells from 1980.

Source: Home Office

From: Social Trends, 1991, Chart 12.20

11.9 Prisoners: by sex and length of sentence
United Kingdom   Numbers and rates

	1984	1985	1986	1987	1988	1989
Males						
Prisoners aged 21 and over serving:						
Up to 18 months	11,797	12,302	11,480	10,771	9,911	9,265
Over 18 months and up to 4 years	8,144	8,878	9,878	10,617	11,308	11,488
Over 4 years less than life	5,426	6,047	6,835	7,989	9,073	10,011
Life sentences	2,361	2,511	2,680	2,852	3,019	3,178
All sentenced male prisoners	27,728	29,738	30,873	32,229	33,311	33,943
Rate per 100,000 male population aged 21 and over	146	155	160	165	170	171
Females						
Prisoners aged 21 and over serving:						
Up to 18 months	616	623	618	576	507	465
Over 18 months and up to 4 years	206	245	306	404	406	381
Over 4 years less than life	71	96	125	172	218	284
Life sentences	63	63	72	77	82	92
All sentenced female prisoners	959	1,027	1,121	1,232	1,212	1,221
Rate per 100,000 female population aged 21 and over	5	5	5	6	6	6

Source: Home Office; Scottish Home and Health Department; Northern Ireland Office

From: Social Trends, 1991, Table 12.21

Definitions and sources

The traffic figures for Great Britain were revised in 1989 (see Road Traffic in Great Britain: Review of Estimates, Transport Statistics Report HMSO 1989). The figures are compiled using roadside traffic counts from which estimates of average daily flow are made. These are combined with information on road lengths to provide estimates of traffic volume in terms of vehicle kilometres.

For other sources see:
Guide to Official Statistics, 1990 edition (200 pages approximately fully indexed) HMSO.

Transport Statistics, Great Britain HMSO.

12.1 Road vehicles in Great Britain: new registration by taxation class

Thousand

| | All vehicles | | | | | | | | Of which body-type cars | | |
| | Private and light goods[1] | | Motor cycles, scooters and mopeds | Goods vehicles[1] | Public transport vehicles | Agricultural tractors[2] | Other vehicles[3] | Total | Total | Percent company | Percent imported |
	Private cars	Other vehicles									
	BMAA	BMAE	BMAD	BMAZ	BMAG	BMAH	BMAY	BMAX	BMAJ	BMAV	BMAC
1985	1 804.0	224.9	125.8	51.8	6.8	40.1	55.4	2 309.3	1 842.1	45	57
1986	1 839.3	231.3	106.4	51.5	8.9	34.8	61.5	2 333.7	1 883.2	46	54
1987	1 962.7	248.3	90.8	54.0	8.7	37.7	70.1	2 473.9	2 016.2	48	50
1988	2 154.7	282.4	90.1	63.4	9.2	45.6	78.6	2 723.5	2 210.3	51	55
1989	2 241.2	293.6	97.3	64.7	8.0	42.5	81.4	2 828.9	2 304.4	51	55
1990	1 942.3	237.6	94.4	44.0	7.4	34.2	78.4	2 438.4	2 005.1	52	56
1989 Dec	87.4	14.7	3.1	2.8	0.5	1.5	5.2	115.0	91.5	57	53
1990 Jan	199.7	23.7	5.2	4.3	0.7	2.5	6.0	242.0	204.4	54	57
Feb	161.6	23.0	5.7	4.1	0.7	2.9	6.3	204.3	166.5	57	55
Mar	202.5	25.8	8.5	4.6	0.8	3.7	8.0	253.9	208.6	54	51
Apr	155.8	20.7	9.2	4.4	0.8	3.7	6.2	201.0	160.7	56	57
May	167.6	20.9	9.7	3.6	0.7	3.3	6.9	212.8	173.2	54	54
Jun	136.8	18.5	8.5	3.7	0.6	2.6	5.9	176.7	141.5	59	54
Jul	43.0	9.1	4.9	1.9	0.3	2.0	3.4	64.7	45.4	64	48
Aug	421.5	34.6	18.0	5.4	0.9	5.7	12.7	498.8	432.6	41	59
Sep	148.8	18.1	9.0	3.9	0.6	2.5	6.3	189.1	153.8	51	54
Oct	125.8	16.5	7.5	3.0	0.5	2.2	6.3	161.9	130.9	55	55
Nov	115.6	16.3	5.2	3.0	0.5	1.9	6.2	148.7	120.6	57	52
Dec	63.6	10.3	2.9	2.1	0.3	1.2	4.1	84.5	66.8	57	50
1991 Jan	159.6	16.4	4.7	2.7	0.6	1.7	5.4	191.2	163.8	57	54
Feb	119.0	14.2	4.3	2.3	0.4	1.8	5.3	147.1	123.3	56	50
Mar	166.0	18.9	7.2	3.0	0.6	2.4	6.7	204.3	171.0	54	54
Apr	116.0	15.6	8.1	2.3	0.5	2.9	6.2	152.0	121.4	60	55

1 For the period up to Oct 1990 retrospective counts within these taxation classes have been estimated. See notes and definitions - Taxation Class Changes.
2 Includes trench diggers, mobile cranes etc but excludes agricultural tractors on exempt licences.
3 Includes crown and exempt vehicles, three-wheelers, pedestrian controlled vehicles, general haulage and showmen's tractors.

Source: Department of Transport

From: Monthly Digest of Statistics, June 1991, Table 13.1

12.2 Household with and without regular use of car   *Great Britain*

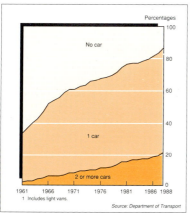

From: Social Trends, 1991, Chart 9.4

12.3 Road and rail passenger transport use   *Great Britain*

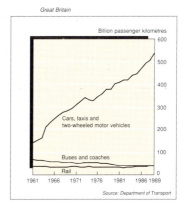

From: Social Trends, 1991, Chart 9.1

TRANSPORT

12.4 Road casualties in Great Britain

Number

	Total casualties		Severity			All severities			
	All ages	Under 15 years	Killed	Seriously injured	Slightly injured	Pedestrians	Pedal cyclists	Motor cyclists and their passengers[1]	Other drivers and their passengers
	BMDA	BMDB	BMDC	BMDD	BMDE	BMDF	BMDG	BMDH	BMDI
1984	324 314	47 187	5 599	73 059	245 656	63 474	30 939	63 821	166 080
1985	317 524	43 644	5 165	70 980	241 379	61 390	26 998	56 591	172 545
1986	321 451	41 426	5 382	68 752	247 317	60 875	26 129	52 280	182 167
1987	311 473	40 013	5 125	64 293	242 055	57 453	26 194	45 801	182 025
1988	322 305	41 050	5 052	63 491	253 762	58 843	25 849	42 836	194 777
1989	341 592	43 041	5 373	63 158	273 061	60 080	28 513	42 630	210 369
1990[2]	340 103	43 500	5 104	60 435	274 563	60 040	26 056	38 840	215 168
1987 Q4	85 022	9 279	1 423	17 363	66 236	16 167	6 280	11 636	50 939
1988 Q1	74 748	8 637	1 104	15 009	58 635	14 796	5 521	8 799	45 632
Q2	75 684	11 057	1 177	15 066	59 441	13 653	6 854	10 925	44 252
Q3	84 988	11 698	1 296	16 609	67 083	14 152	7 345	12 247	51 244
Q4	86 885	9 658	1 475	16 807	68 603	16 242	6 129	10 865	53 649
1989 Q1	77 828	9 364	1 232	14 571	62 025	14 897	5 854	9 027	48 050
Q2	83 305	11 865	1 186	15 324	66 795	14 871	7 660	11 202	49 572
Q3	87 747	12 209	1 422	16 390	69 935	14 126	8 800	12 122	52 699
Q4	92 712	9 603	1 533	16 873	74 306	16 186	6 199	10 279	60 048
1990 Q1[2]	79 787	9 582	1 252	14 444	64 091	15 732	5 532	8 474	50 048
Q2[2]	83 397	12 236	1 225	14 705	67 467	14 500	6 190	10 194	51 623
Q3[2]	86 665	12 475	1 260	15 486	69 919	14 173	7 731	10 894	53 867
Q4[2]	90 254	9 170	1 367	15 800	73 086	15 635	5 713	9 277	59 630

1 Includes riders and passengers of mopeds, motor scooters and combinations.
2 Provisional.

Sources: Department of Transport; Scottish Development Department; Welsh Office

From: Monthly Digest of Statistics, June 1991, Table 13.4

12.5 Local (stage) bus services: fare indices

1985 = 100

	London	English metropolitan areas	English shire counties	England	Scotland	Wales[1]	All Great Britain	All outside London	All outside London and English metropolitan areas
	BAKG	BAKH	BAKI	BAKJ	BAKK	BAKL	BAKM	BAKN	BAKO
1982	98.1	97.6	82.7	89.4	89.1	82.1	88.7	87.7	84.1
1983	100.0	99.5	88.9	94.2	94.6	89.6	93.8	92.8	90.3
1984	91.6	98.6	94.8	95.3	98.4	96.3	95.8	96.6	95.8
1985/86[2]	101.8	100.4	101.2	101.1	100.2	101.1	100.9	100.8	101.0
1986/87[2]	107.8	129.3	106.4	112.7	103.7	106.6	111.1	111.7	105.7
1987/88[2]	113.1	140.0	112.1	119.5	107.8	..	117.4	118.3	110.8
1988/89[2]	125.3	148.9	117.8	127.3	112.2	..	124.6	124.5	116.2
1989/90[2]	138.2	161.2	127.6	138.1	117.8	..	134.5	133.9	124.9
1986 Q3	107.0	128.1	105.5	111.7	103.4	104.7	110.2	110.8	104.9
Q4	107.0	131.0	107.0	113.3	103.7	107.3	111.6	112.5	106.1
1987 Q1	110.2	132.7	108.4	115.1	104.1	109.7	113.2	113.8	107.3
Q2	110.2	135.8	110.3	116.9	106.5	..	115.1	116.1	109.2
Q3	110.2	139.5	111.4	118.5	107.7	..	116.5	117.8	110.3
Q4	110.2	140.8	112.6	119.4	107.8	..	117.3	118.7	111.2
1988 Q1	121.7	144.0	114.1	123.2	109.1	..	120.7	120.6	112.6
Q2	121.7	145.3	115.3	124.3	110.7	..	121.9	121.9	113.9
Q3	121.7	148.3	116.7	125.9	112.1	..	123.4	123.7	115.3
Q4	121.7	150.6	118.5	127.5	112.3	..	124.7	125.3	116.7
1989 Q1	136.1	151.4	120.7	131.6	113.7	..	128.4	127.0	118.7
Q2	136.1	156.2	123.6	134.3	115.8	..	131.0	130.1	121.4
Q3	136.1	161.6	126.5	137.3	117.6	..	133.8	133.5	123.9
Q4	136.1	162.5	128.8	138.7	118.6	..	135.1	134.9	125.9
1990 Q1	144.6	164.5	131.5	142.2	119.4	..	138.2	137.2	128.3
Q2	148.8	168.7	135.8	146.3	122.1	..	142.1	141.0	132.1
Q3	148.8	174.3	138.3	149.1	124.0	..	144.8	144.1	134.4
Q4[3]	148.8	186.0	143.7	154.9	127.7	..	150.2	150.5	139.4
1991 Q1[3]	163.4	189.9	146.9	160.3	128.8	..	155.0	153.5	142.3

1 Figures for Wales since 1986/1987 are omitted because insufficient data are available.
2 Due to rounding financial year data may differ with that published by the Department of Transport.
3 Provisional.

Source: Department of Transport

From: Monthly Digest of Statistics, June 1991, Table 13.6

12.6 Index numbers of road traffic and goods transport by road

Average 1977 = 100

	Index of vehicle kilometres travelled on roads in Great Britain[1]								Index of tonne-kilo-metres of road goods transport[4,5,6]
	Motor traffic								
						Other goods vehicles			
	All motor traffic	Motorcycles etc	Cars and taxis	Buses and coaches	Light vans[2]	Total	Articulated[3]	Pedal cycles	
	BMCA	BMCB	BMCC	BMCD	BMCE	BMCF	BMCG	BMCH	BMCI
1982	115	149	117	109	96	98	108	105	95
1983	117	133	119	115	105	100	113	105	97
1984	123	131	126	119	111	104	119	105	101
1985	125	119	129	113	114	106	121	100	104
1986	132	114	136	114	120	109	124	90	106
1987	141	108	147	126	131	120	144	95	114
1988	151	98	157	133	145	125	158	86	131
1989	163	103	169	139	159	133	174	86	136
1988 Q1	140	78	144	124	137	127	156	75	131
Q2	154	118	160	136	146	124	156	119	132
Q3	162	118	169	149	148	126	156	27	129
Q4	151	85	157	136	146	124	163	75	132
1989 Q1	151	94	155	124	152	133	170	75	138
Q2	167	141	172	149	165	140	177	112	140
Q3	174	156	181	149	163	140	177	134	139
Q4	160	101	166	124	156	135	177	67	138
1990 Q1	156	101	162	124	156	137	171	59	140
Q2	165	141	173	136	159	131	163	66	138
Q3	171	148	179	149	161	131	178	99	137
Q4	160	101	167	124	157	129	171	66	134
1991 Q1	152	86	158	111	149	125	164	53	-

1 All indices have been revised.
2 Not over 30 cwt. unladen weight.
3 Includes vehicles with drawbar trailers.
4 The figures for road goods transport are estimated from a continuing sample enquiry.
5 The quarterly figures relate to 13-week periods and not three calendar months.
6 Revised to exclude estimates of work done by vehicles under 3.5 tonnes gross vehicle weight.

Source: Department of Transport

From: Monthly Digest of Statistics, June 1991, Table 13.3

12.7 British Rail and London Regional Transport railways

Millions

	British Rail: passenger kilometres			London Regional Transport railways: passenger journeys		
	Ordinary fares	Season tickets	Total	Full and reduced fares	Season tickets	Total
	BMGB	BMGD	BMGA	BMGF	BMGG	BMGE
1983	21 757	7 783	29 540	351	212	563
1984	21 818	7 934	29 752	354	295	649
1985	21 585	8 099	29 684	351	373	724
1986	21 948	9 036	30 984	353	413	766
1987	22 607	9 711	32 318	372	431	805
1988	23 276	11 137	34 412	370	451	821
1989	..	..	..	371	438	808
1987 Q2	5 687	2 274	7 961	94	108	202
Q3	6 304	2 290	8 594	100	108	208
Q4	5 676	2 642	8 318	94	108	202
1988 Q1	5 374	2 893	8 267	88	106	195
Q2	5 749	2 711	8 459	87	111	198
Q3	6 318	2 571	8 889	96	112	208
Q4	5 835	2 962	8 797	94	116	210
1989 Q1	5 324	2 853	8 177	85	114	199
Q2[1]	5 459	2 682	8 141	89	95	184
Q3[2]	5 864	2 378	8 242	96	90	187
Q4	5 820	2 852	8 672	100	97	197
1990 Q1	5 282	2 986	8 268	95	102	197
Q2	5 667	2 533	8 200	102	92	194
Q3	6 137	2 391	8 528	106	94	200
Q4	5 623	2 851	8 474	100	98	198
1991 Q1[3]	4 575	2 838	7 413	91	94	185

1 NUR Industrial action on 2 days (BR only).
2 NUR Industrial action on 4 days (BR only).
3 Estimated by DTp.

Source: Department of Transport

From: Monthly Digest of Statistics, June 1991, Table 13.7

12.8 British Rail: freight traffic

| | British Rail [1] | | | | |
| | Freight lifted: million tonnes | | | | |
	Coal and coke	Metals including iron and steel	Other traffic	Total	Net tonne kilometres: millions [2]
	BMHB	BMHC	BMHD	BMHA	BMHE
1982	88.4	14.2	39.2	141.9	15 880
1983	87.9	15.9	41.1	145.1	17 144
1984	25.5	12.4	41.3	78.4	12 720
1985	65.9	14.1	40.5	122.0	15 370
1986	79.7	16.8	43.2	139.6	16 473
1987	77.7	19.1	44.2	141.0	17 297
1988	78.8	20.5	50.1	149.5	17 979
1987 Q2	19.3	5.0	10.9	35.1	4 311
Q3	18.8	4.3	11.3	34.5	4 277
Q4	20.3	5.0	11.3	36.6	4 332
1988 Q1	20.5	5.3	12.4	38.3	4 540
Q2	19.3	5.3	12.1	36.8	4 430
Q3	18.1	4.5	12.8	35.4	4 333
Q4	20.9	5.4	12.8	39.1	4 676
1989 Q1	20.9	5.4	12.5	38.7	4 098
Q2 [3]	18.8	5.2	12.4	36.4	3 754
Q3 [4]	17.3	4.7	11.9	33.9	3 639
Q4	19.5	4.4	12.5	36.4	3 670
1990 Q1	20.2	4.6	12.3	37.1	3 468
Q2	18.6	4.9	12.5	36.0	3 529
Q3	18.1	4.6	11.7	34.0	..
Q4	18.7	4.5	11.0	34.2	..
1991 Q1 [5]	21.1	4.3	11.3	36.7	..

1 Freight train traffic only.
2 Freightliner traffic omitted from 1989 Q1.
3 NUR Industrial action on 2 days.
4 NUR Industrial action on 4 days.
5 Estimated by DTp.

Source: Department of Transport

From: Monthly Digest of Statistics, June 1991, Table 13.8

12.9 UK airlines [1]: aircraft kilometres flown, passengers and cargo uplifted
Tonne-kilometres and seat kilometres used

Monthly averages or calendar months: thousands or tonnes

| | All services | | | Domestic services | | | International services | | |
	Aircraft kilometres flown (000's)	Passengers uplifted (000's)	Cargo uplifted (tonnes) [2]	Aircraft kilometres flown (000's)	Passengers uplifted (000's)	Cargo uplifted (tonnes) [2]	Aircraft kilometres flown (000's)	Passengers uplifted	Cargo uplifted (tonnes) [2]
	BMIA	BMIB	BMIC	BMID	BMIE	BMIF	BMIG	BMIH	BMII
1982	27 579	1 717.4	21 976	4 687	589.0	3 166	22 893	1 128.4	18 810
1983	27 053	1 699.1	24 510	5 058	597.4	3 312	21 995	1 101.7	21 198
1984	29 156	1 879.8	29 958	5 613	694.7	3 784	23 544	1 185.1	26 174
1985	30 955	2 068.7	30 003	5 772	747.7	3 842	25 183	1 321.0	26 161
1986	32 067	2 083.1	31 330	5 932	756.8	3 962	26 136	1 326.3	27 368
1987	33 802	2 374.7	33 780	6 127	837.4	4 235	27 675	1 537.3	29 546
1988	36 562	2 603.7	35 669	6 446	933.2	4 064	30 117	1 670.5	31 606
1989 Jan	37 256	2 335.0	33 839	6 188	778.0	3 341	31 068	1 557.0	30 499
Feb	33 347	2 216.2	35 149	5 629	764.4	3 588	27 718	1 451.7	31 561
Mar	38 050	2 715.4	38 895	6 604	947.6	3 822	31 447	1 767.7	35 073
Apr	38 470	2 777.1	37 177	6 786	980.2	3 625	31 684	1 797.4	33 552
May	41 743	2 974.1	37 457	7 568	1 063.5	4 030	34 176	1 910.6	33 428
Jun	41 935	3 116.7	37 472	7 727	1 105.2	4 125	34 207	2 011.4	33 347
Jul	44 139	3 357.4	38 126	8 108	1 170.1	4 095	36 031	2 187.3	34 032
Aug	43 494	3 325.4	36 094	8 087	1 156.9	4 146	35 407	2 168.5	31 948
Sep	43 270	3 405.2	37 892	7 800	1 196.2	4 085	35 470	2 209.0	33 807
Oct	43 934	3 302.1	40 943	7 627	1 138.5	3 975	36 307	2 163.6	36 968
Nov	40 126	2 817.1	39 550	6 783	967.7	3 940	33 343	1 849.4	35 609
Dec	39 193	2 643.8	39 072	6 270	867.8	3 592	35 923	1 776.0	35 480
1990 Jan	41 256	2 641.3	36 017	6 882	875.8	3 551	34 286	1 765.5	32 466
Feb	37 538	2 556.9	36 871	6 220	848.7	3 514	31 318	1 708.2	33 357
Mar	42 615	3 076.2	43 163	7 154	1 030.3	4 015	35 461	2 045.9	39 149
Apr	43 576	3 278.0	38 874	7 423	1 104.0	3 520	36 153	2 174.0	35 355
May	45 837	3 382.0	40 061	7 643	1 080.1	3 993	38 194	2 237.0	36 070
Jun	45 199	3 510.2	40 534	7 506	1 147.7	4 914	37 694	2 362.5	35 621
Jul	47 397	3 774.2	41 697	7 895	1 213.5	3 783	39 502	2 560.7	37 914
Aug	47 425	3 729.5	39 534	8 106	1 215.4	3 793	39 319	2 512.0	35 742
Sep	46 200	3 679.2	41 657	7 556	1 211.5	3 884	38 644	2 467.7	37 773

1 Scheduled services only. All kilometre statistics are based on standard (Great Circle) distance.
2 Including weight of freight mail, excess baggage and diplomatic bags, but excluding passengers' and crews' permitted baggage.

Source: Civil Aviation Authority

From: Monthly Digest of Statistics, June 1991, Table 13.9

12.10 United Kingdom and Crown Dependency registered trading vessels of 500 gross tons and over [1]
Summary of tonnage by type   *End of year*

	1979	1984	1986		1986	1989
Number				**Number**		
Passenger[3]	101	87	79	Passenger[7]	8	8
Cargo liners	266	97	68	Container (FC)	47	43
Tramps	262	170	132	Other general cargo[6]	128	110
Bulk carriers[2]	203	87	50	Bulk carriers[4]	73	42
Tankers	393	276	169	Tankers	165	136
Container (FC)	80	60	47	Specialised carriers	21	18
				Ro-ro[5]	104	93
All vessels	1 305	777	545	All vessels	546	450
Thousand gross tons				**Thousand gross tons**		
Passenger[3]	606	636	588	Passenger[7]	259	242
Cargo liners	2 248	893	564	Container (FC)	1 369	1 368
Tramps	613	349	244	Other general cargo[6]	510	277
Bulk carriers[2]	6 555	3 398	1 864	Bulk carriers[4]	2 003	1 253
Tankers	13 558	7 463	3 083	Tankers	3 249	2 252
Container (FC)	1 651	1 572	1 369	Specialised carriers	95	122
				Ro-ro[5]	561	510
All vessels	25 232	14 312	7 711	All vessels	8 046	6 025
Thousand deadweight tonnes				**Thousand deadweight tonnes**		
Passenger[3]	177	169	156	Passenger[7]	47	44
Cargo liners	3 011	1 197	748	Container (FC)	1 298	1 308
Tramps	966	560	395	Other general cargo[6]	749	418
Bulk carriers[2]	11 480	6 002	3 321	Bulk carriers[4]	3 569	2 238
Tankers	24 624	12 924	5 499	Tankers	5 908	4 063
Container (FC)	1 623	1 470	1 284	Specialised carriers	122	118
				Ro-ro[5]	380	221
All vessels	41 881	22 322	11 402	All vessels	12 073	8 409

1 In 1986 a new classification of ship types was introduced. This is based on ship descriptions used by Lloyd's Register of Shipping from which the Department of Transport has been taking figures data since 1986. Because of this change, figures for 1986 in Table 10.29 have been given on both bases.
2 Bulk carriers of 10,000 deadweight tonnes and over or approximately 6,000 gross tons and over including combination—ore/oil and ore/bulk/oil-carriers.
3 All vessels with passenger certificates.

4 Bulk carriers (large and small) including combination-ore/oil and ore/bulk/oil-carriers.
5 Ro-ro passenger and cargo vessels.
6 General cargo roll-on/roll-off and lift-on/lift-off vessels, specialised dry cargo vessels and passenger ro-ro vessels.
7 Cruise liner and other passenger.

Source   Department of Transport

From: Annual Abstract of Statistics: 1991, Table 10.29

Definitions and sources

UK food and farming in figures is available free from Ministry of Agriculture, Fisheries and Food, Publications Unit, Lion House, Willowburn Trading Estate, Alnwick, Northumberland NE66 2PF.

For other sources see:
Guide to Official Statistics, 1990 edition (200 pages approximately fully indexed) HMSO.
Farm incomes in the United Kingdom, 1990 edition, HMSO.
Household Food Consumption and Expenditure, 1985, HMSO.

13.1 Agriculture and food in the National Economy

Calendar years

	Average of 1979–81	1986	1987	1988	1989	1990[3]
Agriculture's contribution to Gross Domestic Product [1]						
at current prices (£ million)	4,083	5,576	5,645	5,414	6,217	6,378
at constant 1985 prices (£ million)	4,153	5,026	5,047	5,030	5,276	5,493
% of national GDP (current prices)	2.1	1.8	1.7	1.4	1.5	1.4
Persons engaged in agriculture						
('000 persons)	649	603	589	581	568	561
% of total workforce in employment	2.6	2.5	2.3	2.2	2.2	2.1
Gross fixed capital formation in agriculture						
at current prices (£ million)	1,014	1,049	968	1,125	1,123	..
at constant 1985 prices (£ million)	1,244	1,016	899	965	882	..
% of national GFCF (current prices)	2.5	1.6	1.3	1.2	1.2	..
Imports of food, feed and beverages						(Jan-Sept.)
(£ million)	6,752	10,473	10,658	11,031	11,882	9,589
of which: food, feed and non-alcoholic drinks	6,296	9,507	9,618	9,898	10,658	8,602
alcoholic drinks	456	966	1,040	1,132	1,225	986
Volume index (1985 = 100)	92.6	109.8	111.7	113.4	118.3	121.5
Unit value (price) index (1985 = 100)	73.3	97.1	97.2	99.1	104.1	109.6
% of total UK imports	13.4	12.2	11.3	10.4	9.8	10.1
Exports of food, feed and beverages						(Jan-Sept.)
(£ million)	3,048	5,380	5,592	5,276	6,273	4,731
of which: food, feed and non-alcoholic drinks	2,166	4,072	4,204	3,728	4,505	3,291
alcoholic drinks	881	1,308	1,388	1,548	1,768	1,440
Volume index (1985 = 100)	81.3	113.0	115.1	110.7	123.2	117.8
Unit value (price) index (1985 = 100)	74.9	103.1	103.1	106.2	112.8	120.7
% of total UK exports	6.4	7.4	7.0	6.5	6.7	6.3
UK self-sufficiency in food and feed as a % of:						
all food and feed	57.8	56.5	56.8	53.9	58.5	56.4
indigenous type food and feed	72.8	74.5	72.3	68.9	75.5	73.3
Consumers' expenditure on food and drink						
at current prices (£ million)	38,869	60,233	64,485	71.566	77,867	85,100
of which: household food	23,196	33,070	34,464	36,579	39,181	41,900
meals out [2]	5,749	10,760	12,569	16,233	18,868	21,800
alcoholic drinks	9,924	16,404	17,452	18,754	19,818	21,400
at constant 1985 prices (£ million)	54,104	58,199	59,046	62,011	63,352	63,900
% of total consumers' expenditure	28.1	24.9	24.3	24.0	23.7	23.0
of which: household food	16.8	13.7	13.0	12.2	11.9	11.3
meals out [2]	4.2	4.4	4.7	5.4	5.7	5.9
alcoholic drinks	7.2	6.8	6.6	6.3	6.0	5.8

1 Agriculture is here defined as in the national accounts, that is net of the landlord element and the produce of gardens and allotments.
2 Does not include meals that are included with accommodation.
3 Provisional.

From Agriculture in the UK, 1991

13.2 Economic Indicators for Agriculture

		Income from farming		Cash Flow from farming	
	Net Product[1]	Total income from farming (of farmers, non-principal partners and directors and their spouses and family workers[2]	Farming income (of farmers and spouses[3]	Of farmers, non-principal partners and directors and their spouses and family workers[4]	Of farmers and spouses
1979	2,813	1,511	1,164	1,610	1,263
1980	3,052	1,487	1,065	1,790	1,368
1981	3,566	1,898	1,426	2,272	1,800
1982	4,128	2,330	1,815	2,500	1,985
1983	3,887	1,976	1,420	2,127	1,572
1984	4,816	2,787	2,199	2,923	2,335
1985	4,080	1,808	1,158	2,312	1,662
1986	4,451	2,210	1,525	2,638	1,953
1987	4,486	2,310	1,601	3,072	2,364
1988	4,169	1,874	1,164	2,436	1,726
1989	4,866	2,265	1,513	2,856	2,103
1990 (forecast)	4,954	2,113	1,296	2,748	1,930

1 Net product is a measure of the value added by the agricultural industry to all the goods and services purchased from outside agriculture after provision has been made for depreciation.
2 The return to farmers spouses, non-principal partners and directors for their labour and management skills and on all capital (owned or borrowed) invested in the industry, after providing for depreciation.
3 The return to farmers and their spouses, for their labour,

management skills and own capital invested after providing for depreciation.
4 Cash flow is the pre-tax revenue accruing to farmer and spouse LESS cash outlays (ie spending on material inputs and services and on capital items) in the specific year. The definition has now been extended to include capital grants.

From Agriculture in the UK, 1991

13.3 Agricultural product and input prices (indices 1985 = 100)   Annual averages

	Average[1] 1979-81	1987	1988	1989	1990[2]
Producer prices of agricultural products	82.2	104.2	103.8	111.9	113.6
Crop products	85.9	107.7	101.0	107.9	114.0
of which:					
Cereals	88.7	98.9	96.2	97.2	98.6
Root crops	106.2	152.9	121.0	155.9	170.7
Fresh vegetables	73.8	104.6	100.7	102.4	114.0
Fresh fruit	70.8	108.3	114.2	120.2	137.2
Animals and animal products	80.4	102.2	105.5	114.2	113.3
of which:					
Animals for slaughter	81.2	100.4	103.3	112.0	107.4
Milk	76.1	107.5	115.2	122.7	125.6
Eggs	94.1	91.4	77.5	93.2	106.0
Wool	91.1	99.3	98.9	98.0	97.9
Agricultural input prices	74.3	101.2	106.1	112.1	116.8
Inputs currently consumed in agriculture	75.3	98.8	103.3	109.0	113.2
of which:					
Animal feedingstuffs	82.0	101.3	106.9	112.7	114.0
Seeds	99.1	100.6	100.7	96.7	97.5
Fertilisers and soil improvers	77.2	82.9	85.4	92.4	94.0
Energy & lubricants	58.5	77.7	73.7	78.0	90.2
Capital inputs	70.5	113.9	120.8	128.9	135.7
of which:					
Machinery & other equipment	71.7	110.1	116.1	122.9	128.9
Buildings	68.5	119.1	127.2	137.0	145.0
Index of ratio of product to input price	110.6	103.0	97.8	99.8	97.3
Labour costs	66.2	110.2	115.4	124.5	140.4
Estimated rates of interest on bank advances to agriculture (per cent)	16.8	12.1	12.5	16.2	17.1

1 1980 based indices re-referenced to 1985 = 100.
2 Forecast.

From Farm Incomes in the UK, 1991

13.4 Agricultural land area in the United Kingdom ('000 hectares)   At June each year

	Average of 1979-81	1986	1987	1988	1989	1990[5]
Total area of agricultural land[1]	18,948	18,676	18,622	18,575	18,553	18,525
This comprises:						
Total cereals	3,877	4,024	3,935	3,896	3,873	3,693
of which:						
Wheat	1,357	1,997	1,994	1,886	2,083	2,042
Barley	2,343	1,916	1,830	1,878	1,652	1,522
Oats	155	97	99	120	119	106
Rye and mixed corn	23	14	13	12	12	13
Triticale[2]	. .	. .	. .	. .	8	9
Other arable crops (excluding potatoes)	588	824	958	969	882	989
of which:						
Oilseed rape	77	299	388	347	321	397
Sugar beet	212	201	200	198	197	197
Hops	6	4	4	4	4	4
Peas for harvesting dry and field beans	78	150	208	260	215	222
Other crops	216	165	156	156	146	169
Potatoes	208	178	177	180	175	178
Horticulture	246	213	199	209	208	203
of which:						
Vegetables grown in the open	166	146	132	141	141	138
Orchard fruit	48	38	37	37	36	34
Soft fruit	18	15	15	15	15	15
Ornamentals[3]	12	12	12	13	14	13
Other	2	2	2	2	2	2
Bare fallow	67	48	42	58	65	62
All grass under five years old	1,986	1,723	1,691	1,613	1,534	1,587
All grasses five years old and over	5,132	5,077	5,112	5,161	5,236	5,250
Rough grazing:						
Sole right	5,151	4,829	4,743	4,712	4,710	4,669
Common (estimated)	1,211	1,216	1,216	1,236	1,236	1,236
All other land (including woodland on agricultural holdings)[4]	481	543	547	562	620	658

1 The data cover all holdings, including minor holdings, in England and Wales but exclude minor holdings in Scotland and Northern Ireland.
2 Collected separately for the first time in 1989 (Great Britain only).
3 Hardy nursery stock, bulbs and flowers.
4 In Great Britain other land comprises farm roads, yards, buildings (excluding glasshouses), ponds and derelict land. In Northern Ireland other land includes land under bog, water, roads, buildings etc and wasteland not used for agriculture.
5 Provisional.

From Farm Incomes, 1991

AGRICULTURE

13.5 Livestock numbers in the United Kingdom ('000)

At June each year

	Average 1979–81	1987	1988	1989	1990[1]
Total cattle & calves	13,384	12,158	11,872	11,977	12,108
of which:					
Dairy cows	3,237	3,042	2,911	2,865	2,854
Beef cows	1,481	1,343	1,373	1,495	1,601
Heifers in calf	855	774	834	793	753
Total sheep & lambs	31,163	38,701	40,942	42,967	44,217
of which:					
Ewes and shearlings	14,924	18,068	19,017	19,994	20,500
Other sheep	16,239	20,663	21,925	22,973	23,717
Total pigs	7,836	7,942	7,980	7,509	7,606
of which:					
Sows in pig and other sows for breeding	730	713	703	660	664
Gilts in pig	110	107	101	97	110
Total fowls	125,712	128,628	130,809	120,198	124,763
of which:					
Table fowls (including broilers)	58,300	70,754	75,305	70,042	73,356
Laying fowls	46,201	38,498	37,389	33,957	34,152
Growing pullets	14,727	12,230	11,236	9,411	10,329

From Farm Incomes, 1991

13.6 Landings of fish of British taking: landed weight and value

United Kingdom

	Landed weight (Thousand tonnes)							Value (£ thousand)						
	1983	1984	1985	1986	1987	1988	1989	1983	1984	1985	1986	1987	1988	1989
Total all fish	748.4	733.7	762.1	716.9	790.4	742.0	671.8	280,263	297,863	323,825	361,680	435,162	402,893	394,346
Total wet fish	676.2	661.0	687.2	629.3	679.3	645.5	580.4	229,233	240,587	258,904	284,161	339,223	310,412	299,354
Demersal:														
Catfish	1.3	1.3	1.4	1.5	1.6	1.6	1.8	494	589	755	1,035	1,318	1,223	1,557
Cod	112.6	90.9	90.0	76.5	93.4	7.7	68.2	72,466	65,258	69,945	69,140	86,312	74,849	69,568
Dogfish	10.6	12.3	13.8	11.6	13.6	13.0	11.3	3,287	3,547	4,550	6,376	7,201	7,066	7,762
Haddock	122.6	107.5	132.2	131.0	102.4	97.6	71.9	57,361	64,204	67,757	79,019	78,079	68,708	61,918
Hake	2.3	2.5	2.7	3.0	3.3	3.6	4.0	2,519	2,794	3,851	4,549	5,540	6,389	7,699
Halibut	0.2	0.1	0.1	0.1	0.1	0.1	0.1	360	319	365	372	446	482	442
Lemon sole	5.9	5.7	5.7	5.0	5.3	5.4	5.0	5,493	5,972	7,137	8,265	9,018	8,788	8,945
Plaice	21.0	21.5	20.4	21.3	25.7	27.3	26.2	13,774	14,097	13,821	16,048	22,541	22,348	20,400
Redfish	0.3	0.4	0.2	0.1	0.2	0.2	0.2	73	98	75	56	62	71	101
Saithe (Codfish)	12.8	12.1	14.4	17.7	15.2	14.4	11.8	3,520	2,759	3,874	6,361	7,077	5,657	4,868
Skate and ray	6.5	6.8	7.0	6.9	8.6	8.4	7.4	2,881	2,792	2,992	3,908	4,742	4,796	4,596
Sole	2.3	2.4	2.7	3.2	3.1	3.0	2.8	5,877	6,717	9,093	14,500	16,991	14,067	14,370
Turbot	0.5	0.5	0.5	0.6	0.7	0.7	0.5	1,426	1,660	1,976	2,607	3,612	3,690	3,014
Whiting	57.0	60.7	50.2	41.1	51.9	45.6	38.4	17,715	23,006	18,997	18,460	25,198	21,143	20,834
Livers[1]	..	..	..	..	..	..				1		1		
Roes	0.5	0.6	0.5	0.5	0.5	0.5	0.4	275	321	390	348	499	473	457
Other demersal	71.1	69.7	60.5	63.1	55.0	66.9	63.0	14,563	17,580	21,596	25,660	35,690	37,198	41,861
Total	427.5	395.0	402.3	383.2	381.3	366.0	313.1	202,084	211,713	227,175	256,705	304,326	276,948	268,392
Pelagic:														
Herring	54.2	71.5	95.4	106.0	100.3	93.2	99.4	6,994	9,076	11,493	11,881	12,213	11,244	11,257
Mackerel[2]	175.5	186.3	174.2	132.0	189.4	176.1	157.9	18,649	18,768	18,694	14,766	21,484	20,791	18,702
Other pelagic	19.0	8.2	15.4	7.9	8.3	10.2	10.0	1,506	1,030	1,543	809	1,200	1,429	1,003
Total	248.7	266.0	285.0	246.1	298.0	279.5	267.4	27,149	28,874	31,370	27,456	34,897	33,464	30,962
Total shell fish	72.2	72.6	74.8	87.6	111.1	96.5	91.3	51,030	57,274	64,920	77,519	95,939	92,481	94,992
Cockles	5.8	5.4	7.8	19.4	39.0	24.6	14.8	327	311	476	1,165	4,285	3,276	1,274
Crab	11.3	14.0	13.5	12.6	13.5	15.2	14.1	6,373	8,422	8,557	9,599	11,400	13,908	14,143
Lobster	1.0	1.2	1.1	1.0	1.1	1.3	1.3	5,748	7,696	8,030	7,680	9,346	10,163	10,650
Mussels	5.9	4.3	5.8	6.3	4.9	6.9	9.0	468	314	467	541	543	906	1,190
Nephrop (Norway lobster)	22.1	22.2	24.8	25.4	24.2	27.4	27.0	24,061	24,545	31,548	39,119	43,007	44,585	42,122
Oysters[3]	0.3	0.4	0.5	0.6	0.1	0.1	0.5	636	770	734	845	241	225	179
Shrimps[4]	1.1	0.7	0.8	1.4	3.3	1.7	1.8	739	566	711	1,254	2,346	1,785	2,627
Whelks	1.3	2.2	1.6	2.0	2.7	2.0	1.2	226	456	355	418	665	473	251
Other shell fish	23.4	22.2	18.9	18.9	22.3	17.3	22.0	12,452	14,194	14,042	16,898	24,106	17,165	22,556

1 Including the raw equivalent of any liver oils landed.
2 Includes transhipments of mackerel or herring ie caught by British vessels but not actually landed at British ports. These quantities are transhipped to foreign vessels and are later recorded as exports.
3 The weight of oysters is calculated on the basis of one tonne being equal to 15,748 oysters in England and Wales.
4 From 1986 data from prawn is also included.

From: Annual Abstract of Statistics 1991 Table 9.13

13.7 Consumers' expenditure in the United Kingdom

	1979		1984		1989	
	£m	%	£m	%	£m	%
Household expenditure on food	20,988	17.6	29,304	14.7	39,220	12.1
Expenditure on meals out	4,978	4.2	8,337	4.2	18,646	5.7
Total expenditure on food	25,966	21.7	37,641	18.9	57,866	17.8
Alcoholic drink	8,665	7.3	14,316	7.2	20,065	6.2
Total food and drink	34,631	29.0	51,957	26.2	77,931	24.0
Total consumers' expenditure	119,516	100.0	198,895	100.0	324,348	100.0

From National Food Survey, 1988

13.8 Consumption and expenditure for main food groups
Per person per week

		Consumption			Expenditure		
		1987	1988	1989	1987	1988	1989
		(ounces) ([1])			(pence)		
Milk and cream	(PT or Eq)	4.07	4.01	3.93	108.01	112.46	120.85
Cheese		4.09	4.13	4.07	34.11	38.21	40.01
Meat and meat products		36.97	36.59	35.94	292.01	303.65	327.91
Fish		5.09	5.06	5.20	53.31	57.63	63.09
Eggs	(no)	2.89	2.67	2.29	20.60	19.32	17.75
Fats and oils		10.04	9.86	9.48	33.37	34.51	36.10
Sugar and preserves		9.35	8.79	8.22	17.37	17.77	17.65
Fruit and vegetables		114.80	115.01	114.41	212.74	226.99	243.07
Cereals (inc. bread)		54.87	54.12	53.33	167.87	180.17	191.86
Beverages		2.70	2.66	2.60	41.85	42.34	43.19
Other foods		—	—	—	40.57	43.63	48.20
All foods		—	—	—	£10.22	£10.77	£11.50

(1) except where otherwise stated

From National Food Survey, 1988

13.9 Contributions of selected foods to energy intake
kcals per person per day

	1987	1988	1989
Dairy products	275	271	264
of which:			
Whole milk	158	146	133
Meat and meat products	321	314	311
Total fats	299	291	277
of which:			
Butter	64	60	52
Margarine	119	113	104
Other fats	117	118	121
Packet sugar	120	112	103
Potatoes	88	85	83
Cereals	630	617	605
of which:			
White bread	146	142	140
Other bread	139	140	132
Other cereal products	345	335	332

From National Food Survey, 1988

Definitions and sources

All tables and charts in this section are taken from *Social Trends* each edition of which has different coverage so that the same tables do not necessarily appear year to year.

For other sources see:

Guide to Official Statistics, 1990 edition (200 pages approximately fully indexed) HMSO.

14.1 Leisure time in a typical week: by sex and employment status, 1989

Great Britain

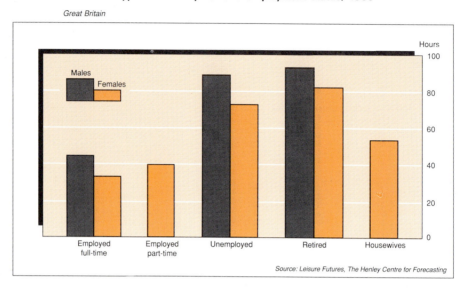

Source: Leisure Futures, The Henley Centre for Forecasting

From: Social Trends 1991, Chart 10.1

14.2 Participation in selected home-based leisure activities: by age: 1987

Great Britain

Percentages and numbers

	Age							
	16–19	20–24	25–29	30–44	45–59	60–69	70 and over	All ages
Percentage in each group participating in each activity in the 4 weeks before interview								
Watching TV	99	99	99	99	99	98	97	99
Visiting/entertaining friends or relations	97	97	98	96	95	93	92	95
Listening to radio	94	94	92	90	87	84	75	88
Listening to records/tapes	97	94	89	84	71	58	33	73
Reading books	59	56	60	64	60	62	55	60
Gardening	17	23	37	53	57	57	40	46
DIY	28	39	51	56	48	39	18	43
Dressmaking/needlework/ knitting	14	21	25	29	30	32	23	27
Sample size (= 100%) (numbers)	1,384	1,821	1,827	5,162	4,140	2,654	2,541	19,529

Source: General Household Survey

From: Social Trends 1991, Table 10.4

14.3 Reading of national newspapers: by sex and by age, 1971 and 1989

Great Britain

	Percentage of adults reading each paper in 1989			Percentage of each age group reading each paper in 1989				Readership[1] (millions)		Readers per copy (numbers)
	Males	Females	All adults	15–24	25–44	45–64	65 and over	1971	1989	1989
Daily newspapers										
The Sun	26	22	24	31	25	22	17	8.5	10.8	2.6
Daily Mirror	22	17	20	21	18	23	19	13.8	8.8	2.8
Daily Mail	10	9	10	8	9	11	10	4.8	4.3	2.5
Daily Express	9	8	9	7	7	11	10	9.7	3.9	2.5
Daily Star	7	4	6	8	7	5	3	.	2.7	2.9
The Daily Telegraph	6	5	6	3	4	8	7	3.6	2.5	2.3
The Guardian	4	2	3	3	4	3	1	1.1	1.3	2.9
Today	5	3	4	5	5	3	2	.	1.8	3.2
The Times	3	2	2	3	3	2	2	1.1	1.1	2.5
The Independent	3	2	3	3	3	2	1	.	1.2	2.8
Financial Times	2	1	2	1	2	2	—	0.7	0.7	3.5
Any daily newspaper [2]	69	62	65	65	64	69	64	. .	. .	. .
Sunday newspapers										
News of the World	30	27	29	37	31	27	20	15.8	13.0	2.5
Sunday Mirror	22	19	20	23	20	21	17	13.5	9.2	3.1
The People	18	16	17	15	16	19	17	14.4	7.5	2.8
Sunday Express	12	11	11	9	9	15	15	10.4	5.2	2.7
The Mail on Sunday	12	12	12	14	14	12	7	.	5.5	2.8
The Sunday Times	9	7	8	8	10	9	4	3.7	3.6	2.8
Sunday Telegraph	5	4	4	3	4	6	5	2.1	2.0	3.1
The Observer	6	4	5	5	6	5	3	2.4	2.2	3.3
Any Sunday newspaper [3]	75	70	72	74	72	75	68	. .	. .	. .

1 Defined as the average issue readership and represents the number of people who claim to have read or looked at one or more copies of a given publication during a period equal to the interval at which the publication appears.

2 Includes the above newspapers plus the Daily Record.
3 Includes the above newspapers plus the Sunday Post and Sunday Mail

Source: National Readership Surveys, Joint Industry Committee for National Readership Surveys; Circulation Review, Audit Bureau of Circulation

From: Social Trends 1991, Table 10.9

14.4 Television viewing[1] and radio listening: by age
United Kingdom

Hours and minutes and percentages

	Television viewing					Radio listening				
	1985	1986	1987	1988	1989	1985	1986	1987	1988	1989
Age groups (hours:mins per week)										
4 – 15 years	19:59	20:35	19:14	18:34	18:27	2:24	2:12	2:07	2:13	2:21
16 – 34 years	21:36	21:10	20:03	20:36	20:34	11:42	11:24	11:18	11:40	12:07
35 – 64 years	28:04	27:49	27:25	27:17	26:07	9:43	9:56	10:16	10:33	11:10
65 years and over	36:35	36:55	37:41	37:25	36:29	8:04	8:27	8:44	8:49	9:00
All aged 4 years and over	26:33	25:54	25:25	25:21	24:44	8:40	8:40	8:52	9:12	9:46
Reach[2] (percentages)										
Daily	79	78	76	77	78	43	43	43	43	44
Weekly	94	94	93	94	94	78	75	74	73	74

1 Viewing of live television broadcasts from the BBC, ITV and Channel 4.
2 Percentage of UK population aged 4 and over who viewed TV for at least three consecutive minutes or listened to radio for at least half a programme a day.

Source: Broadcasters' Audience Research Board; Audits of Great Britain; British Broadcasting Corporation;

From: Social Trends 1991, Table 10.6

LEISURE

14.5 Leisure based consumer durables: by household type, 1988−89

Great Britain   Percentages

	1 adult aged 16−59	2 adults aged 16−59	Small family[1]	Large family[2]	Large adult house-hold[3]	2 adults 1 or both aged 60 and over	1 adult aged 60 and over	All house-holds
Percentage of households with								
Television	91	98	99	99	100	99	96	98
Video cassette recorder	38	72	74	80	77	34	8	53
Home computer	9	15	35	44	29	4	1	18

1 One or 2 persons aged 16 and over and 1 or 2 persons aged under 16.
2 One or more persons aged 16 and over and 3 or more persons aged under 16, or 3 or more persons aged 16 and over and 2 persons aged under 16.
3 Three or more persons aged 16 and over, with or without 1 person aged under 16.

Source: General Household Survey

From: Social Trends 1991, Table 10.5

14.6 Trade deliveries of LPs, cassettes, CD's and singles[1]

United Kingdom

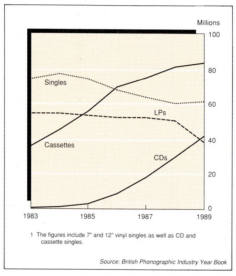

1 The figures include 7" and 12" vinyl singles as well as CD and cassette singles.

Source: British Phonographic Industry Year Book

From: Social Trends 1991, Chart 10.8

14.7 Spectator attendance[1] at selected sporting events

Thousands

	1971/72	1981/82	1989/90
Football League (England & Wales)	28,700	20,006	19,466
Greyhound racing	8,800	6,100	5,400
Horse racing	4,200	3,700	4,924
Scottish Football League	4,521	2,961	3,575
Rugby Football Union (England)	700	750[6]	2,250
Motor sports[2]	. .	1,300	1,650
Rugby Football League[3]	1,170	1,226	1,689
Test and County cricket	984	994	751[7]
English basketball	2	85	203
Motorcycle sports[4]	. .	20	25
Scottish basketball[5]	9	14	9

1 Estimated.
2 Car and kart racing only.
3 League matches only.
4 Excluding speedway.
5 National league and cup matches only.
6 1982 season.
7 1988 season.

Source: Organisations concerned

From: Social Trends, 1991, Table 10.18

14.8 Membership of selected organisations for young people

United Kingdom   Thousands

Membership	1971	1981	1989
Cub Scouts	265	309	359[8]
Brownie Guides	376	427	385[9]
Scouts[1]	215	234	200
Girl Guides[2]	316	348	241
Sea Cadet Corps	18	19	16
Army Cadet Force	39	46	39
Air Training Corps	33	35	38
Combined Cadet Force	45	44	42
Boys Brigade[3]	140	154	110
Girls Brigade[3]	97	94	91
Methodist Association of Youth Clubs	115	127	90
National Association of Boys Clubs	164	186	200
Youth Clubs UK — Boys	179	430[7]	319
— Girls	140	341[7]	242
National Federation of Young Farmers Clubs[4]			
— Males	24	28	21
— Females	16	23	17
Young Men's Christian Association[5]			
Registered members			
— Males	35	36	886
— Females	13	19	730
Registered participants[6]	154	151	1,616
Duke of Edinburgh's Award			
Participants	122	170	200
Awards gained			
Bronze	18	23	22
Silver	7	10	9
Gold	3	5	4

1 Includes Venture Scouts (15½-20 years).
2 Includes Ranger Guides (14-18 years) and Young Leaders (15-18 years). In addition to the United Kingdom figures also include the Channel Islands and the Isle of Man and British Guides in Foreign Countries.
3 Figures relate to British Isles.
4 Figures relate to England, Wales, and the Channel Islands and to young people aged between 10 and 25 in 1971, and between 10 and 26 in 1981.
5 Figures relate to persons aged under 21.
6 The 1989 figure is not comparable with earlier years because of a change in definition.
7 Figures include membership of clubs affiliated to four local associations.
8 Includes Beaver Scouts.
9 Includes Rainbow Guides (4 or 5-7 years).

Source: Organisations concerned

From: Social Trends 1991, Table 11.9

14.9 Domestic holidays taken by United Kingdom residents: by region 1989

United Kingdom

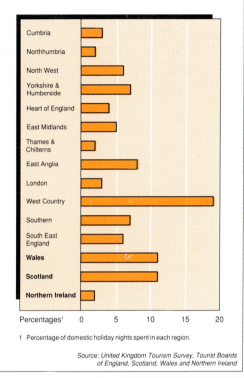

1 Percentage of domestic holiday nights spent in each region.

Source: United Kingdom Tourism Survey, Tourist Boards of England, Scotland, Wales and Northern Ireland

From: Social Trends 1991, Chart 10.22

Definitions and sources

Educational establishments in the United Kingdom may be administered and financed in one of three different ways:

Public sector: by local education authorities, which form part of the structure of local government;

Assisted: by governing bodies which have a substantial degree of autonomy from public authorities but which receive grants direct from central government sources;

Independent: by the private sector, including individuals, companies, and charitable institutions.

The pupil/teacher ratios used here are the ratios of all pupils to all teachers employed on the day of annual count (with due allowance for part-timers). The count is taken in January (September in Scotland from 1974–75).

Ages are measured at 31 August for 1980–81 onwards (except in Northern Ireland; age detail have been estimated).

Further education
The term 'further education' may be used in a general sense to cover all education after the period of compulsory education. More commonly it excludes those staying on at secondary school, and those studying at universities.

Higher education
The term 'higher education' as used here covers all advanced courses (including teacher training courses) in universities and institutions of higher and further education, that is those leading to qualifications above General Certificate of Education 'A' level, Scottish Certificate of Education 'H' grade, BTEC National Diploma and Ordinary National Diploma; or their equivalents.

There are very helpful explanatory notes given in Social Trends appendix Part 3, as well as in the publications of the education departments (summarised in bulletins).

University students are counted on 31 December. All other students are counted on 1 November in England and Wales, and in October in Scotland and Northern Ireland.

Full-time includes sandwich courses; part-time includes evening only courses.

For other sources see:

Guide to Official Statistics, 1990 edition (200 pages approximately fully indexed) HMSO.

Education Statistics for the United Kingdom, 1990 edition (80 pages approximately fully indexed) HMSO.

15.1 School pupils [1]: by type of school [2]

United Kingdom			Thousands	
	1961	1971	1981[3]	1989[3]
Public sector schools (full and part-time)				
Nursery schools	31	50	89	100
Primary schools				
Under fives	} 4,906 {	301	459	768
Other primary[4]		5,601	4,712	4,024
Secondary schools				
Under school leaving age	..	..	4,202	3,146
Over school leaving age	..	..	404	406
All secondary schools	3,165	3,555	4,606	3,551
Total public sector	8,102	9,507	9,866	8,443
Independent schools	680	621	619	641
Special schools (full-time equivalent)	77	103	147	118
All schools	8,859	10,230	10,632	9,203

1 Part-time pupils are counted as one (except for special schools).
2 See Appendix, Part 3: Main categories of educational establishments and Stages of education.
3 Figures for Scottish components are at previous September.
4 In Scotland, 11 year olds are customarily in primary schools.

Source: Department of Education and Science; Scottish Education Department; Welsh Office; Northern Ireland Department of Education

From Social Trends, 1991, Table 3.5

15.2 Pupils in public sector secondary education[1,2] by type of school

England, Wales, Scotland and Northern Ireland
Percentages and thousands

	1971	1981	1989
England *(percentages)*			
Maintained secondary schools			
Middle deemed secondary	*1.9*	*7.0*	*6.3*
Modern	*38.0*	*6.0*	*3.9*
Grammar	*18.4*	*3.4*	*3.4*
Technical	*1.3*	*0.3*	*0.1*
Comprehensive	*34.4*	*82.5*	*85.9*
Other	*6.0*	*0.9*	*0.4*
Total pupils (= 100%) (thousands)	2,953	3,840	2,945
Wales *(percentages)*			
Maintained secondary schools			
Middle deemed secondary	*0.1*	*0.1*	—
Modern	*22.3*	*1.8*	*0.2*
Grammar	*15.4*	*1.3*	*0.2*
Comprehensive	*58.5*	*96.6*	*98.9*
Other	*3.7*	*0.3*	*0.7*
Total pupils (= 100%) (thousands)	191	240	192
Scotland *(percentages)*			
Public sector secondary schools			
Selective	*28.3*	*0.1*	—
Comprehensive	*58.7*	*96.0*	*100.0*
Part comprehensive/part selective	*13.0*	*3.8*	—
Total pupils (= 100%) (thousands)	314	408	312
Northern Ireland *(percentages)*			
Public sector secondary schools			
Secondary intermediate	*87.7*	*88.6*	*87.8*
Grammar	*11.8*	*11.4*	*12.2*
Technical intermediate	*0.5*	—	—
Total pupils (= 100%) (thousands)	96	119	103

1 See Appendix, Part 3: Main categories of educational establishments and Stages of education.
2 Counts at January except 1981 data for Scotland which are at the preceding September.

Source: Education Statistics for the United Kingdom, *Department of Education and Science*

From Social Trends, 1991, Table 3.8

15.3 Class sizes as taught[1]: by type of school

England

	1977	1981	1986	1989
Primary schools				
Percentage of classes taught by:				
One teacher in classes with				
1—20 pupils	*16*	*20*	*17*	*14*
21—30 pupils	*46*	*55*	*58*	*62*
31 or more pupils	*34*	*22*	*19*	*17*
Two or more teachers	*4*	*3*	*6*	*7*
Average size of class (numbers)	28	26	26	26
Number of classes (thousands)	171	161	142	145
Secondary schools				
Percentage of classes taught by:				
One teacher in classes with				
1—20 pupils	*42*	*44*	*46*	*47*
21—30 pupils	*42*	*45*	*45*	*44*
31 or more pupils	*12*	*8*	*6*	*4*
Two or more teachers	*4*	*2*	*3*	*4*
Average size of class (numbers)	22	22	21	21
Number of classes (thousands)	165	174	155	137

1 Class size related to one selected period in each public sector school on the day of the count in January. Middle schools are either primary or secondary for this table - see Appendix, Part 3: Stages of education.

Source: Department of Education and Science

From Social Trends, 1991, Table 3.9

15.4 Pupil/teacher ratios[1]: by type of school

United Kingdom			Ratio
	1971	1981	1989
Public sector schools			
Nursery	26.6	21.5	21.6
Primary	27.1	22.3	21.9
Secondary	17.8	16.4	15.0
All public sector schools	23.2	19.0	18.3
Non-maintained schools	14.0[2]	13.1[3]	11.3
Special schools	10.5[2]	7.4	6.1
All schools	22.0[2]	18.2[3]	17.1

1 See Appendix, Part 3: Pupil/teacher ratios.
2 Excludes independent schools in Scotland.
3 Excludes independent schools in Northern Ireland.

Source: Education Statistics for the United Kingdom, *Department of Education and Science*

From Social Trends, 1991, Table 3.33

EDUCATION

15.5 Selected statistics of manpower employed in education: by type of establishment

United Kingdom   Thousands and percentages

	1970/71	1975/76	1980/81	1985/86[1]	1987/88[2]	Percentage who were graduates 1980/81	Percentage who were graduates 1987/88
Full-time teachers and lecturers							
Schools							
Public sector							
Primary schools[3]	203	240	222	201	206	*16*	*29*
Secondary schools	199	259	281	266	253	*54*	*63*
Non-maintained schools[4]	36	39	43	43	43	*63*	*71*
Special schools	10	17	19	19	20	*22*	*33*
Total	448	555	565	529	522	*39*	*49*
Establishments of further education[5]	69	86	89	93	94	*43*	*47*
Universities[6]	29	32	34	31	31	*99*	*99*
Total educational establishments[7]	546	677	693	657	651	*42*	*51*

1 Data for 1984/85 have been used for manpower in schools in Scotland.
2 Data for maintained schools in England and Wales are estimated.
3 Includes nursery schools.
4 Excludes independent schools in Scotland and Northern Ireland.
5 Includes former colleges of education.
6 Excludes Open University. There were 663 professors and lecturers and 5,544 part-time tutorial and counselling staff employed by the Open University at January 1988. Also excludes the independent University College of Buckingham.
7 Includes miscellaneous teachers in England and Wales (3.7 thousand in 1987/88) not shown elsewhere above.

Source: Education Statistics for the United Kingdom, Department of Education and Science

From Social Trends, 1991, Table 3.31

15.6 Percentage of school leavers with grades A—C at GCSE[1]: by subject and sex, 1987/88
Great Britain

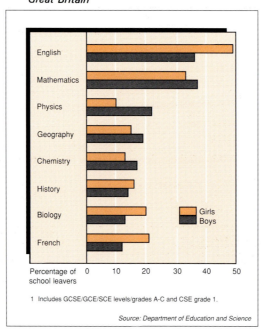

1 Includes GCSE/GCE/SCE levels/grades A-C and CSE grade 1.

Source: Department of Education and Science

From: Social Trends, 1991, Chart 3.14

15.7 Higher education[1] — part-time students[2]: by sex, type of establishment and age

United Kingdom   Thousands and percentages

	Males					Females				
	1970 /71	1975 /76	1980 /81	1985 /86	1988 /89	1970 /71	1975 /76	1980 /81	1985 /86	1988 /89
Part-time students by establishment										
Universities	18.1	19.3	22.6	26.3	29.0	5.7	7.0	10.7	16.0	21.1
Open University[3]	14.3	33.6	37.6	41.7	45.0	5.3	22.0	30.1	36.0	40.3
Polytechnics and colleges										
— part-time day courses	69.8	80.2	110.5	112.2	118.5	6.7	15.4	30.8	49.9	67.0
— evening only courses	39.8	35.0	35.1	34.4	38.1	5.0	5.8	15.2	20.3	26.5
Total part-time students	142.0	168.1	205.7	214.6	230.6	22.7	50.2	86.8	122.2	154.9
Part-time students by age *(percentages)*										
18 years and under	..	..	*6*	*4*	*4*	..	..	*4*	*2*	*2*
19 – 20 years	..	..	*16*	*14*	*12*	..	..	*9*	*7*	*7*
21 – 24 years	..	..	*23*	*22*	*19*	..	..	*18*	*17*	*16*
25 years and over	..	..	*54*	*60*	*65*	..	..	*69*	*73*	*75*

1 See Appendix, Part 3: Stages of education.
2 Excludes students enrolled on nursing and paramedic courses at Department of Health establishments; 1988/89 data is not yet available but there were some 94 thousand in 1987/88.
3 Calendar years beginning in second year shown. Excludes short course students up to 1982/83. Data for 1988/89 are not available, but in 1987/88 there were 10.4 thousand specialised short course students for whom data by sex were not available; these have been excluded.

Source: Education Statistics for the United Kingdom, *Department of Education and Science*

From Social Trends, 1991, Table 3.23

15.8 Higher education[1]— full-time students: by sex, origin and age

United Kingdom   Thousands and percentages

	Males					Females				
	1970 /71	1975 /76	1980 /81	1985 /86	1988 /89	1970 /71	1975 /76	1980 /81	1985 /86	1988 /89
Full-time students by origin										
From the United Kingdom										
Universities[2] — post-graduate	23.9	23.2	20.7	21.0	21.1	8.0	10.2	11.3	12.6	14.0
— first degree	128.3	130.1	145.1	134.3	138.4	57.0	73.6	96.2	99.9	108.4
— other[3]				1.5	1.3				1.2	1.3
Polytechnics and colleges	102.0	109.3	111.9	143.5	147.9	113.1	120.1	96.4	132.2	146.7
Total full-time UK students	254.2	262.6	277.7	300.4	308.6	178.2	203.8	203.9	245.9	270.5
From abroad	20.0	38.6	40.7	38.4	42.5	4.4	9.9	12.6	15.3	22.6
Total full-time students	274.2	301.2	318.4	338.7	351.1	182.6	213.7	216.5	261.3	293.1
Full-time students by age *(percentages)*										
18 years and under	*10*	*11* [4]	*16*	*15*	*15*	*17*	*14* [4]	*19*	*17*	*16*
19 – 20 years	*36*	*35*	*37*	*38*	*37*	*45*	*42*	*41*	*42*	*39*
21 – 24 years	*38*	*36*	*30*	*29*	*31*	*24*	*28*	*25*	*26*	*28*
25 years and over	*15*	*19*	*17*	*18*	*18*	*14*	*16*	*15*	*15*	*18*

1 See Appendix, Part 3: Stages of education.
2 Origin is on fee-paying status except for EC students domiciled outside the United Kingdom who from 1980/81 are charged home rates but are included with students from abroad. From 1984 origin is based on students' usual places of domicile.
3 University first diplomas and certificates.
4 In 1980 measurement by age changed from 31 December to 31 August.

Source: Education Statistics for the United Kingdom, *Department of Education and Science*

From: Social Trends 1991, Table 3.19

EDUCATION

15.9 Higher education qualifications obtained[1]: by type of qualification and sex

United Kingdom   Thousands

Type of qualification	1981	1986	1987
Below degree level[2]			
Males	45	54	53
Females	17	26	34
Total	62	80	87
First degree[3]			
Males	76	78	79
Females	48	61	64
Total	124	139	143
Post-graduate[4]			
Males	24	27	29
Females	13	16	17
Total	37	42	46
All higher education qualifications			
Males	144	158	162
Females	78	103	115
Total	222	262	277

1 Includes estimates of successful completions of public sector professional courses (43 thousand in 1987-88). Excludes successful completions of nursing and paramedical courses at Department of Health establishments (38 thousand in 1987-88) and the private sector.
2 First university diplomas and certificates; CNAA diplomas and certificates below degree level; BTEC/SCOTVEC higher diplomas and certificates.
3 University degrees, and estimates of CNAA degrees (and equivalent) and university validated degrees (Great Britain only).
4 Universities, CNAA and PGCEs.

Source: Education Statistics for the United Kingdom, Department of Education and Science

From Social Trends, 1990, Table 3.25

15.10 Student awards — real value and parental contributions

England & Wales

	Standard maintenance grant[1] (£)	Index of the real value of the grant deflated by		Average assessed contribution by parents[4] (percentages)
		Retail prices index[2]	Average earnings index[3]	
1979/80	1,245	100	100	13
1980/81	1,430	99	99	13
1981/82	1,535	96	96	14
1982/83	1,595	93	91	19
1983/84	1,660	92	89	20
1984/85	1,775	94	89	25
1985/86	1,830	91	87	30
1986/87	1,901	92	82	30
1987/88	1,972	91	78	31
1988/89	2,050	90	74	31
1989/90	2,155	88	..	..

1 Excludes those studying in London and those studying elsewhere living in the parental home. Prior to 1982/83 Oxford and Cambridge were also excluded. Since 1984/85 has included an additional travel allowance of £50.
2 September 1979 = 100.
3 Great Britain average earnings for the whole economy has been used as the deflator. February 1980 = 100.
4 Assuming full payment of parental and other contributions including a notional assessment in respect of students for whom fees only were paid by LEAs. Of the students assessed for parental contributions in 1988/89 there were 103.2 thousand mandatory award holders (29 per cent) who were receiving the maximum grant because their parent's assessed contribution was nil.

Source: Department of Education and Science; Department of Employment

From Social Trends, 1991, Table 3.27

15.11 Destination of first degree graduates
Great Britain

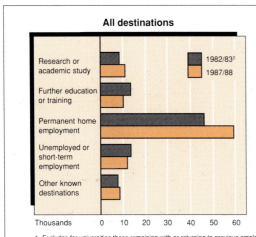

All destinations

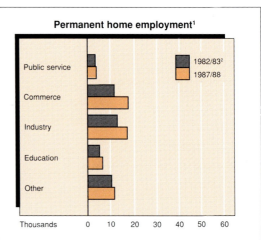

Permanent home employment[1]

1 Excludes for universities those remaining with or returning to previous employer.
2 Excludes one polytechnic which did not supply the informaton for 1982/83.

Source: Department of Education and Science

From: Social Trends, 1991, Chart 3.26

Key Data 91, © Crown copyright 1991

15.12 Government expenditure on education: by type

United Kingdom £ million and percentages

	1970 −71	1980 −81	1988 −89
Current expenditure (£ million)			
Schools			
Nursery	7	52	5,290
Primary	546	2,840	
Secondary	619	3,695	6,474
Special	50	443	888
Further and adult education[1]	265	1,591	3,187
Training of teachers: tuition	60	78	197
Universities[1]	246	1,264	1,966
Other education expenditure	95	516	1,003
Related education expenditure	443	1,693	2,402
Total current expenditure	2,331	12,172	21,406
Capital expenditure (£ million)			
Schools	245	480	477
Other education expenditure	164	288	435
Total capital expenditure	409	768	911
Total government expenditure (£ million)	2,740	12,940	22,317
Of which, expenditure by local authorities	2,318	11,160	19,123
Expenditure as a percentage of GDP	5.1	5.5	..

1 Includes tuition fees.

Source: Department of Education and Science; Central Statistical Office

From Social Trends, 1991, Table 3.34

Definitions and sources

Further health statistics at regional level, including manpower, are given in *Regional Trends*.
For other sources see:
Guide to Official Statistics, 1990 edition (200 pages approximately fully indexed) HMSO.

Health and Personal Social Services Statistics for England HMSO.
Social Security Statistics HMSO.

16.1 Expectation of life: by sex and age

United Kingdom   Years

	Males			Females		
	1901	1989	2001	1901	1989	2001
Expectation of life[1]						
At birth	45.5	72.8	73.9	49.0	78.4	79.8
At age:						
1 year	53.6	72.4	73.4	55.8	78.0	79.3
10 years	50.4	63.7	64.6	52.7	69.2	70.5
20 years	41.7	53.9	54.8	44.1	59.4	60.6
30 years	33.8	44.3	45.1	35.9	49.5	50.8
40 years	26.1	34.8	35.7	28.3	39.8	41.0
50 years	19.3	25.6	26.7	21.1	30.4	31.6
60 years	13.3	17.4	18.5	14.6	21.7	22.7
70 years	8.5	10.9	11.9	9.2	14.2	15.1
80 years	4.9	6.2	6.9	5.3	8.2	8.8

1 Further number of years which a person could be expected to live. See Appendix, Part 7: Expectation of life.

Source: Government Actuary's Department

From: Social Trends, 1991, Table 7.2

16.2 National Health Service hospital summary: all specialties

United Kingdom

	1971	1976	1981	1984	1985	1986	1987–88	1988–89
All in-patients								
Discharges and deaths (thousands)	6,437	6,525	7,179	7,666	7,884	7,959	8,088	8,144
Average number of beds available daily[1] (thousands)	526	484	450	429	421	409	392	373
Average number of beds occupied daily (thousands)								
Maternities	19	16	15	14	13	13	13	. .
Other patients	417	378	350	333	327	316	304	. .
Total—average number of beds occupied daily	436	394	366	347	341	330	317	. .
Patients treated per bed available (number)	12.3	13.6	16.0	17.8	18.7	19.4	20.6	21.8
Average length of stay (days)								
Medical patients	14.7[2]	12.1	10.2[2]	9.1	8.7	8.5	. .	. .
Surgical patients	9.1[2]	8.6	7.6[2]	6.9	6.7	6.5	. .	. .
Maternities	7.0[2]	6.7	5.6[2]	4.9	4.7	4.5	. .	. .
Percentage of live births in hospital[2]	89.8	97.6	98.9	99.0	99.1	99.1	99.1	99.0
Day case attendances (thousands)	. .	565[2]	863	1,081	1,166	1,288	1,207	1,259
New out-patients[3] (thousands)								
Accidents and emergency	9,358	10,463	11,342	12,279	12,492	12,682	12,797	13,201
Other out-patients	9,572	9,170	9,816	10,376	10,604	10,758	10,350	10,409
Average attendances per new patient (numbers)[4]								
Accidents and emergency	1.6	1.6	1.4	1.4	1.3	1.3	1.3	1.3
Other out-patients	4.2	4.0	4.4	4.3	4.3	4.3	4.2	4.2

1 Staffed beds only.
2 Great Britain only.
3 The 1971 and 1976 figures for out-patients in Scotland include ancillary departments.
4 Patients attending out-patients clinics in England solely for attention of a minor nature and not seen by a doctor, eg to have a dressing changed are no longer counted.

Source: Department of Health; Scottish Health Service, Common Services Agency; Welsh Office; Department of Health and Social Services, Northern Ireland

From: Social Trends, 1991, Table 7.31

HEALTH AND SOCIAL SECURITY

16.3 NHS hospital in-patient waiting lists[1]:
by specialty

United Kingdom   Thousands

Specialty	1976	1981	1986	1989	1990
General surgery	200.5[2]	169.1[2]	180.3[2]	173.0[2]	172.5
Orthopaedics	109.8	145.1	160.5	153.0	155.9
Ear, nose or throat	121.7	115.4	132.2	123.2	127.5
Gynaecology	91.8	105.6	106.6	96.4	98.0
Oral surgery	26.5	35.5	56.3	50.4	53.7
Plastic surgery	44.7	49.2	46.1	46.8	45.3
Ophthalmology	41.2	43.4	64.6	88.9	92.7
Urology	22.0[3]	29.1[3]	42.7	47.1	47.2
Other	42.5	44.2	41.3	48.0	48.8
All specialties	700.8	736.6	830.6	827.0	841.6

1 At 30 September each year except 1990, at 31 March.
2 Includes the Northern Ireland figures for 'Urology'.
3 Great Britain only.

Source: Department of Health; Welsh Office;
Scottish Health Service, Common Services Agency;
Department of Health and Social Services, Northern Ireland

From Social Trends, 1991, Table 7.30

16.4 Family practitioner and dental services

United Kingdom

	General medical and pharmaceutical services					General dental services		
	Number of doctors[1] in practice (thousands)	Average number of patients per doctor (thousands)	Prescriptions dispensed[2] (millions)	Average cost[3] per prescription (£)	Average number of prescriptions per person	Average prescription cost[4] per person (£)	Number of dentists[5] in practice (thousands)	Average number of persons per dentist (thousands)	Average number of courses of treatment per dentist (thousands)
---	---	---	---	---	---	---	---	---	---
1961	23.6	2.25	233.2	0.41	4.7	1.9	11.9	4.4	1.4
1971	24.0	2.39	304.5	0.77	5.6	4.3	12.5	4.5	2.0
1981	27.5	2.15	370.0	3.46	6.6	23.0	15.2	3.7	2.2
1985	29.7	2.01	393.1	4.77	7.0	33.4	17.0	3.3	2.2
1986	30.2	1.99	397.5	5.11	7.0	36.0	17.3	3.3	2.2
1987	30.7	1.97	413.6	5.47	7.3	40.0	17.6	3.2	2.2
1988	31.2	1.94	427.7	5.91	7.5	44.1	18.0	3.2	2.2
1989	31.5	1.91	435.8	6.26	7.5	47.2	18.4	3.1	2.1

1 Unrestricted principals only. See Appendix, Part 7: Unrestricted principals.
2 Prescriptions dispensed by general practitioners are excluded. The number of such prescriptions in the United Kingdom is not known

precisely, but in Great Britain during 1988 totalled some 32.1 million.
3 Average total cost per prescription.
4 Total cost including dispensing fees and cost.
5 Principals plus assistants.

Source: Department of Health

From Social Trends, 1991, Table 7.33

HEALTH AND SOCIAL SECURITY

16.5 Selected causes of death[1]: by sex and age, 1951 and 1989

United Kingdom
Rates per 1,000 deaths

	Males					Females				
	Under 15 years	15-34	35-44	45-64	65 and over	Under 15 years	15-34	35-44	45-59	60 and over
1951										
Infectious diseases	71.6	205.8	153.2	65.2	13.9	92.5	333.2	128.6	41.9	6.9
Cancer[2]	29.7	109.7	206.3	231.3	141.1	30.5	132.5	315.5	329.6	134.1
Circulatory diseases[3]	6.7	97.4	205.7	306.1	419.8	10.2	125.4	174.7	229.6	425.2
Respiratory diseases[4]	149.2	56.9	94.8	166.6	157.9	165.2	67.3	73.9	90.4	134.8
Accidents and violence	103.9	335.9	141.7	45.0	18.7	74.4	76.1	53.0	35.7	21.0
All other diseases	638.8	194.4	198.3	185.9	248.5	627.2	265.4	254.3	272.8	278.1
1989										
Infectious diseases	51.6	12.5	11.1	5.2	3.2	47.7	23.8	8.4	6.6	3.7
Cancer[2]	79.1	113.2	222.8	334.7	258.1	79.8	251.1	536.8	534.2	212.4
Circulatory diseases[3]	29.4	69.6	298.6	468.8	478.3	31.2	83.0	145.4	243.8	495.4
Respiratory diseases[4]	81.6	33.6	38.9	61.3	142.3	85.4	48.4	32.4	58.5	123.7
Accidents and violence	190.4	614.1	265.2	46.9	13.5	143.0	369.9	121.8	44.8	15.9
All other diseases	567.9	157.0	163.3	83.0	104.6	612.9	223.8	155.1	112.0	148.9

1 See Appendix; Part 7: Death certificates.
2 The figures for neoplasms include both malignant and benign cancers.
3 Includes heart attacks and strokes.

4 In 1984 the coding procedure was changed, reducing the number of deaths assigned to respiratory causes.

Source: Office of Population Censuses and Surveys; General Register Offices for Scotland and Northern Ireland

From: Social Trends 1991, Table 7.5

16.6 Reported AIDS cases: EC comparison, 1990[1]

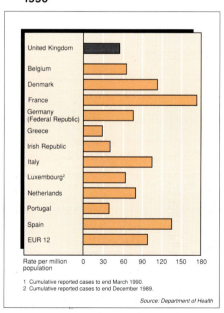

1 Cumulative reported cases to end March 1990.
2 Cumulative reported cases to end December 1989.

Source: Department of Health

From: Social Trends 1991, Chart 7.9

16.7   Cervical cancer: deaths and screening

Great Britain
Thousands and percentages

	1976	1981	1983	1984	1985	1986	1987–88	1988–89
Deaths	2.4	2.2	2.2	2.1	2.2	2.2	2.1	2.1
Smears taken	2,923	3,442	3,669	3,911	4,455	4,468	4,754	5,032
Smears as a percentage of women aged 15 and over	*13.3*	*15.2*	*16.0*	*17.0*	*19.3*	*19.2*	*20.4*	*21.5*

Source: Department of Health; Scottish Health Service, Common Services Agency; Welsh Office

From: Social Trends 1991, Table 7.25

HEALTH AND SOCIAL SECURITY

16.8 Children on child protection registers: by reason, 1989[1]
England & Wales

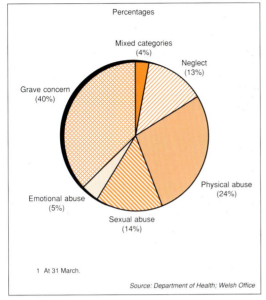

1 At 31 March.

Source: Department of Health; Welsh Office

From: Social Trends 1991, Chart 7.36

16.9 Elderly people in residential accommodation[1]
United Kingdom

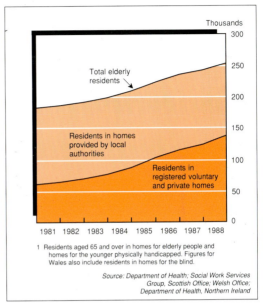

1 Residents aged 65 and over in homes for elderly people and homes for the younger physically handicapped. Figures for Wales also include residents in homes for the blind.

Source: Department of Health; Social Work Services Group, Scottish Office; Welsh Office; Department of Health, Northern Ireland

From: Social Trends 1991, Chart 7.37

16.10 Government expenditure on the national health service
Years ended 31 March

£ million

	1979/80	1980/81	1981/82	1982/83	1983/84	1984/85	1985/86	1986/87	1987/88	1988/89[3]	1989/90
Current expenditure											
Central government:											
Hospitals and Community health services[1]:											
Running expenses[2]	6 168	8 162	9 033	9 691	10 448	10 912	11 570	12 356	13 644	15 171	16 676
Family practitioner services:											
General medical services	570	754	866	973	1 049	1 192	1 291	1 373	1 514	1 697	1 860
Pharmaceutical services[2]	986	1 213	1 394	1 599	1 767	1 919	2 041	2 235	2 459	2 797	3 083
General dental services[2]	400	494	562	629	681	747	776	877	972	1 126	1 191
General ophthalmic services[2]	107	123	148	238	191	206	170	155	184	204	137
Administration	365	450	474	488	452	473	475	553	627	691	736
less Payments by patients:											
Hospital services	– 42	– 57	– 69	– 72	– 81	– 84	– 92	– 99	– 106	– 103	– 115
Pharmaceutical services	– 49	– 88	– 107	– 125	– 134	– 149	– 158	– 204	– 256	– 202	– 242
Dental services	– 78	– 106	– 132	– 163	– 179	– 197	– 225	– 261	– 290	– 282	– 340
Ophthalmic services	– 33	– 34	– 38	– 45	– 51	– 52	– 14	– 1	– 1	–	
Total	– 202	– 285	– 346	– 405	– 445	– 482	– 489	– 565	– 653	– 587	– 697
Departmental administration	88	109	121	109	135	137	142	171	193	212	247
Other services	191	236	183	206	219	218	283	324	336	326	471
Total current expenditure	8 673	11 256	12 435	13 528	14 497	15 322	16 259	17 479	19 276	21 637	23 704
Capital expenditure											
Central government	522	688	832	857	886	990	1 085	1 142	1 209	1 299	1 460
Total expenditure											
Central government	9 195	11 944	13 267	14 385	15 383	16 312	17 344	18 621	20 485	22 936	25 164

1 Including the school health service.
2 Before deducting payments by patients.

3 Provisional.

Source Central Statistical Office

From: Annual Abstract of Statistics, 1991, Table 3.3

Definitions and sources

The 10-yearly population census is also known as the Census of Population and Housing. It is the main source of information on housing conditions throughout the country for the local level.

The main source of national trend data on the state of the environment in the UK is the annual *Digest of Environment Protection and Water Statistics.* Definitional notes and guidance on further sources are given as footnotes to the tables.

For other sources see:

Guide to Official Statistics, 1990 edition (200 pages approximately fully indexed) HMSO

Housing and construction Statistics, HMSO.

Digest of Environmental Protection and Water Statistics, HMSO.

17.1 Stock of dwellings: by tenure[1]
United Kingdom

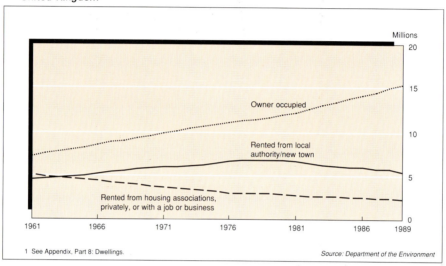

1 See Appendix, Part 8: Dwellings.

Source: Department of the Environment

From: Social Trends 1991, Chart 8.1

17.2 Tenure: by socio-economic group of head of household, 1988
Great Britain

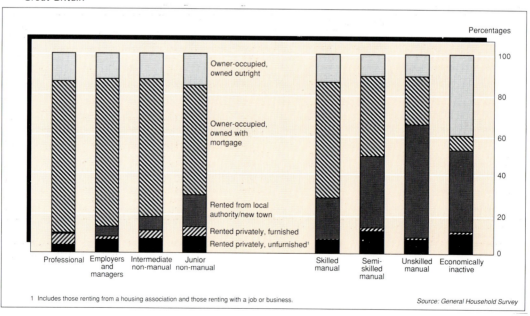

1 Includes those renting from a housing association and those renting with a job or business.

Source: General Household Survey

From: Social Trends 1991, Chart 8.25

HOUSING AND ENVIRONMENT

17.3 Sales of dwellings owned by local authorities and new towns[1]
England & Wales

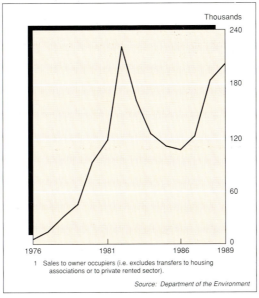

1 Sales to owner occupiers (i.e. excludes transfers to housing associations or to private rented sector).

Source: Department of the Environment

From: Social Trends 1991, Chart 8.4

17.4 Building societies mortgages[1]: balances, arrears and possessions

United Kingdom

	All mortgages			Loans in arrears at end-period (thousands)		Properties taken into possession in period (thousands)
	Number of mortgages (thousands)	Balance due on mortgages (£ million)	Average mortgage balance (£)	By 6—12 months	By over 12 months	
1971	3,896	10,332	2,652	..	..	..
1976	4,609	22,565	4,896	..	..	..
1981	5,490	48,875	8,903	..	..	4.2 [2]
1982	5,645	56,596	10,044	23.8	4.8	6.0
1983	5,928	67,474	11,382	25.6	6.5	7.3
1984	6,314	81,882	12,968	41.9	8.3	10.9
1985	6,657	96,775	14,498	49.6	11.4	16.8
1986	7,023	115,699	16,474	45.2	11.3	20.9
1987	7,182	130,870	18,222	48.2	13.0	22.9
1988	7,369	153,167	20,785	37.2	8.9	16.1
1989	7,929	..	..	58.0	12.0	13.7
1990	8,062	..	..	76.3	18.8	14.4

1 Council of Mortgage Lenders estimates as at 31 December in each
  year except 1990, 30 June.
2 Change in method of estimation.

Source: Council of Mortgage Lenders

From Social Trends, 1991, Table 8.23

HOUSING AND ENVIRONMENT

17.5 Indicators of fixed investment in dwellings

| | Fixed investment in dwellings (£ million, 1985 prices) | Orders received by contractors for new houses (GB) (£ million, 1985 prices) | Housing starts (GB)+ | | | Housing completions (GB)+ | | | Building societies | | Average price of new dwellings: mortgages approved[1,2] (NSA) |
			Private enterprise (thousands)	Housing associations (thousands)	Local authorities new towns and government departments (thousands)	Private enterprise (thousands)	Housing associations (thousands)	Local authorities new towns and government departments (thousands)	Commitments on new dwellings (£ million, current prices)	Advances on new dwellings (£ million, current prices)	
	DFEG	FCAS	FCAT	CTOQ	CTOU	FCAV	CTOS	CTOW	AHLO	AHLS	FCAY
1984	12 550	4 293	158.3	12.6	27.3	159.1	16.6	33.9	2 980	2 900	34 160
1985	11 854	5 290	165.7	12.5	21.9	156.4	13.1	27.2	2 963	2 900	37 304
1986	12 805	5 642	180.6	12.9	20.4	169.8	12.5	22.8	3 626	3 517	43 647
1987	13 705	6 010	194.5	12.4	20.0	179.4	12.0	20.0	3 702	3 488	51 290
1988	14 873†	6 143	219.8	13.8	16.6	193.5	11.9	19.7	4 831	4 696	64 615
1989	14 420	4 689	168.4	13.7	14.3	173.4	11.9	16.5	..	..	74 976
1990	12 549	3 349	133.0	16.6	8.7	153.1	14.6	16.4	..	..	78 917
1985 Q4	2 951	1 439	43.9	3.3	5.4	39.3	3.0	5.8	850	777	39 458
1986 Q1	3 031	1 290	41.6	3.3	4.7	39.6	3.3	5.9	819	802	40 470
Q2	3 040	1 460	44.8	3.4	4.6	41.1	3.2	5.9	972	869	43 443
Q3	3 250	1 441	47.8	3.2	5.7	42.9	2.8	5.2	962	920	45 065
Q4	3 484	1 450	46.5	3.2	5.2	46.1	3.2	5.8	873	927	45 749
1987 Q1	3 552	1 537	46.9	2.5	5.2	43.1	2.5	5.0	839	852	47 460
Q2	3 215	1 448	46.5	3.4	5.4	45.3	3.4	5.0	820	831	49 604
Q3	3 553	1 501	50.5	3.4	4.6	45.5	3.3	4.9	950	871†	52 428
Q4	3 385	1 525	50.6	3.2	4.7	45.4	2.8	5.2	1 092†	934†	55 474
1988 Q1	3 488†	1 557	53.5	3.2	4.4	48.6	3.0	4.9	1 304	1 117	60 395
Q2	3 947	1 472	55.6	3.0	4.3	48.2	2.9	5.5	1 312	1 199	63 342
Q3	3 801	1 523	53.1	3.9	3.9	50.4	2.9	5.0	1 192	1 277	67 425
Q4	3 637	1 592	57.6	3.7	4.0	46.3	3.1	4.3	1 023	1 104	69 342
1989 Q1	4 139	1 370	49.5	4.2	3.5	46.6	3.4	4.5	1 012	1 024	71 828
Q2	3 371	1 202	44.4	3.5	3.8	43.5	2.6	4.0	1 162	1 086	75 213
Q3	3 366	1 024	38.9	2.8	4.0	42.5	3.1	4.2	1 162	998	75 993
Q4	3 544	1 093	35.6	3.2	3.0	40.8	2.8	3.8	1 144	1 113	76 308
1990 Q1	3 251	995	35.8	4.0	2.9	38.4	2.8	4.6	1 097	1 067	78 536
Q2	3 209	847	32.0	4.2	2.3	37.6	3.5	4.0	853	926	79 558
Q3	3 045	780	32.1	4.5	2.0	38.8	3.8	4.5	922	892	79 777
Q4	3 044	727	33.1	3.9	1.5	38.3	4.5	3.3	1 030	889	77 886
1991 Q1	2 778	800†	32.4†	3.9†	1.4†	36.5†	4.6†	3.3†	945	909	..
1989 Jun	..	377	14.2	1.2	1.2	13.8	1.1	1.2	384†	350†	76 201
Jul	..	335	13.2	1.0	1.8	14.7	0.9	1.4	333	304	76 729
Aug	..	341	13.0	0.9	1.2	13.3	1.2	1.2	399	331	75 714
Sep	..	348	12.7	0.9	1.0	14.5	1.0	1.6	430	363	75 639
Oct	..	364	13.3	1.2	1.0	14.1	0.9	1.1	389	364	75 023
Nov	..	367	11.6	1.0	1.0	14.0	1.0	1.1	374	369	75 785
Dec	..	362	10.7	1.0	1.0	12.7	0.9	1.6	381	380	78 844
1990 Jan	..	363	12.0	1.4	1.0	13.1	1.1	1.5	373	349	78 632
Feb	..	316	11.3	1.1	0.9	12.2	0.9	1.3	387	363	78 805
Mar	..	316	12.5	1.5	1.0	13.1	0.8	1.8	337	355	78 225
Apr	..	288	11.4	1.4	0.8	12.3	1.1	1.3	282	323	79 764
May	..	275	10.1	1.6	0.7	12.3	1.0	1.4	292	302	79 745
Jun	..	284	10.5	1.2	0.8	13.0	1.4	1.3	279	301	79 176
Jul	..	268	10.2	1.3	0.8	12.5	1.3	1.3	296	312	80 404
Aug	..	275	11.3	1.7	0.8	13.5	1.5	1.6	325	294	80 127
Sep	..	237	10.6	1.5	0.4	12.8	1.0	1.6	301	286	78 651
Oct	..	233	11.9	1.3	0.5	13.3	1.8	1.1	337	307	76 359
Nov	..	209	10.6	1.2	0.5	13.2	1.3	1.1	332	294	77 094
Dec	..	286	10.6	1.4	0.5	11.8	1.4	1.1	361	288	81 060
1991 Jan	..	301	11.4†	0.9†	0.6†	12.0†	1.5	1.1	340	311	75 031
Feb	..	258	9.7	0.9	0.5	11.3	1.6	1.1†	308	312	74 716
Mar	..	240	11.3	2.1	0.3	13.2	1.5†	1.1	297	286	76 917
Apr	..	247†	11.7	1.1	0.2	12.3	1.4	0.8	351	329	..

1 Mortgages with building societies by private owners. The series covers only dwellings on which building societies have approved mortgages during the period. The cost of land is included.
2 The Abbey National ceased to operate as a building society in July 1989, but to ensure continuity in the data its results are included in the building society sector whenever possible.

Sources: Central Statistical Office;
Department of the Environment;
Scottish Development Department;
Building Societies Association

From: Economic Trends, June 1991, Table 12

HOUSING AND ENVIRONMENT

17.6 Emissions of sulphur dioxide and nitrogen oxides[1]: EC comparison

Thousand tonnes

	Sulphur dioxide			Nitrogen oxides[1]		
	1975	1981	1987	1975	1981	1987
United Kingdom	5,310	4,387	3,863	2,365	2,328	2,429
Belgium	..	856[3]	610[4]	..	317[3]	271[4]
Denmark	418	363	248	182	212	266
France	3,329	2,735	1,517	1,608	1,779	1,652
Germany (Fed. Rep.)	3,325	3,034	2,223[5]	2,532	2,851	2,969[5]
Greece	..	546	..	..	217	..
Irish Republic	186	189	138[6]	60	68	68[6]
Italy	3,250[2]	3,211	2,075[5]	1,499[2]	1,585	1,570[5]
Luxembourg	..	24	13[6]	..	23	22[6]
Netherlands	386	445	274[5]	447	547	560[5]
Portugal	178	266	286[6]	104	166	303[6]
Spain	..	2,543	..	..	937	..

1 Nitrogen oxides expressed as nitrogen dioxide equivalent.
2 1976.
3 1980.
4 1983.
5 1986.
6 1985.

Source: Department of the Environment

From: Social Trends, 1991, Table 9.21

17.7 Acidity of rain and wet deposited acidity 1989
United Kingdom

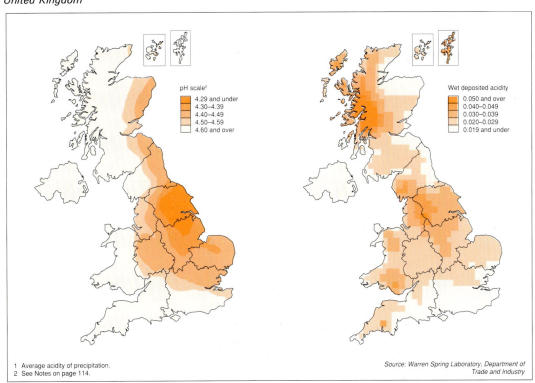

1 Average acidity of precipitation.
2 See Notes on page 114.

Source: Warren Spring Laboratory, Department of Trade and Industry

From: Regional Trends, 1991, Chart 6.15

17.8 Quality of rivers and canals: by regional water authority, 1988 [1]

Percentages and kilometres

	Good quality	Fair quality	Poor quality	Bad quality	Total length (= 100%) (kilo-metres)
North West	57	23	16	4	5,900
Northumbrian	88	10	2	0	2,785
Severn Trent	53	37	9	1	6,156
Yorkshire	73	14	11	2	6,034
Anglian	57	34	8	0	4,453
Thames	66	29	5	0	2,418
Southern	75	19	6	0	2,161
Wessex	60	35	4	1	2,466
South West	54	34	11	1	2,601
Welsh	80	13	6	1	4,802
England & Wales	65	24	9	2	39,776

1 In some cases estimates of river quality have been based on financial years and/or more than one year's data.

Source: Water Authorities

From: Social Trends 1991, Table 9.28

17.9 Forest crown density
United Kingdom

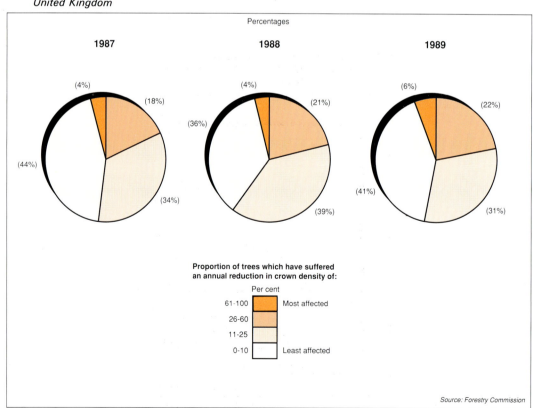

Percentages

1987 (4%) (18%) (44%) (34%)

1988 (4%) (21%) (36%) (39%)

1989 (6%) (22%) (41%) (31%)

Proportion of trees which have suffered an annual reduction in crown density of:

Per cent
- 61-100 — Most affected
- 26-60
- 11-25
- 0-10 — Least affected

Source: Forestry Commission

From: Social Trends 1991, Chart 9.38

HOUSING AND ENVIRONMENT

17.10 Recycled scrap as a proportion of total consumption for selected materials
United Kingdom

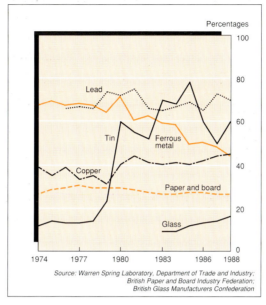

Source: Warren Spring Laboratory, Department of Trade and Industry;
British Paper and Board Industry Federation;
British Glass Manufacturers Confederation

From: Social Trends, 1991, Chart 9.35

17.11 Noise - complaints received by Environmental Health Officers: by source[1]
England & Wales

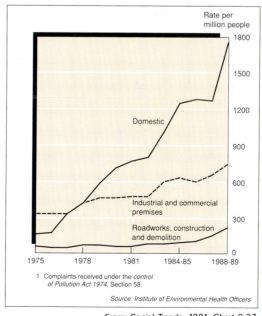

1 Complaints received under the *control of Pollution Act 1974*, Section 58.

Source: Institute of Environmental Health Officers

From: Social Trends, 1991, Chart 9.37

Definitions and sources

All the data shown (except for the chart below) are first published in *Energy Trends* (Monthly).

Detailed definitions and sources are given in the supplementary notes to the *Monthly Digest of Statistics* and in the *Digest of United Kingdom Energy Statistics* (Annual).

For other sources see:

Guide to Official Statistics, 1990 edition (200 pages approximately fully indexed) HMSO.

Digest of UK Energy Statistics, HMSO.

Energy Trends.

Department of Energy.

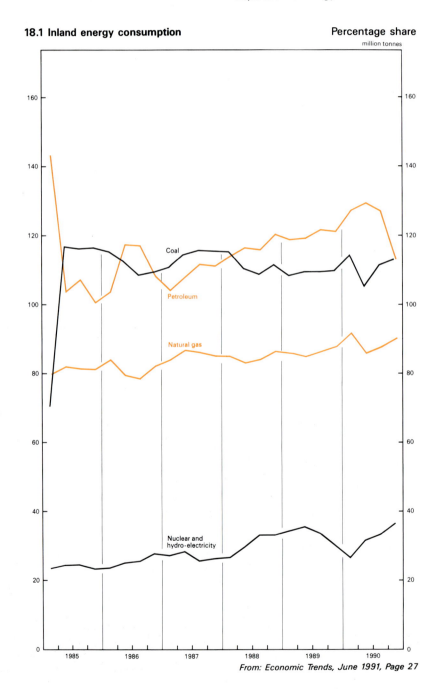

18.1 Inland energy consumption

Percentage share

From: *Economic Trends, June 1991, Page 27*

18.2 Inland energy consumption: primary fuel input basis

Million tonnes of oil or oil equivalent

| | Not seasonally adjusted | | | | | | | Seasonally adjusted (annual rates)[7] | | | | | | |
| | | | | Primary electricity | | | | | | | Primary electricity | | | |
	Coal[1]	Petroleum[2]	Natural gas[3]	Nuclear	Natural flow hydro[5]	Net imports	Total	Coal[1,4]	Petroleum[2,4]	Natural gas[3,4]	Nuclear	Natural flow hydro[5]	Net imports[6]	Total
	BHBB	BHBC	BHBD	BHBE	BHBF	BHBM	BHBA	BHBH	BHBI	BHBJ	BHBK	BHBL	BHBN	BHBG
1986	66.8	66.2	49.2	12.6	1.4	1.0	197.2	65.1	65.1	48.0	12.6	1.4	1.0	193.1
1987	68.3	64.3	50.5	11.7	1.2	2.8	198.9	67.2	63.5	49.6	11.7	1.2	2.8	196.1
1988	65.9	68.3	47.9	13.5	1.4	3.1	200.1	65.9	68.3	49.7	13.5	1.4	3.1	201.9
1989	63.6	69.5	47.4	15.4	1.4	3.0	200.2	64.3	70.4	50.5	15.4	1.4	3.0	204.9
1990[8]	64.0	71.3	48.7	14.2	1.5	2.9	202.4	65.2	72.8	51.5	14.2	1.5	2.9	208.1
1990 Feb	5.8	5.8	5.4	1.2	0.2	-	18.4	65.6	75.4	53.0	15.0	1.4	-	210.4
Mar*	7.0	7.2	5.9	1.3	0.2	-	21.7	68.9	75.0	53.4	13.5	1.5	-	212.4
Apr	4.7	5.6	4.2	1.1	0.1	0.3	16.0	60.5	73.9	48.7	14.3	1.5	3.7	202.6
May*	4.3	5.4	2.6	1.1	0.1	0.2	13.7	62.8	75.5	50.2	13.6	1.5	3.0	206.6
Jun*	5.2	7.3	2.9	1.1	0.1	0.3	16.9	61.7	78.3	51.3	12.7	1.5	3.5	209.0
Jul	4.0	5.7	1.9	0.9	0.1	0.3	12.8	60.0	81.4	49.6	13.3	1.5	4.1	209.9
Aug	4.1	5.4	1.8	1.0	0.1	0.3	12.7	68.5	75.9	51.5	14.8	1.5	3.7	215.9
Sep*	5.7	6.5	3.0	1.3	0.1	0.4	17.0	67.0	65.6	51.9	14.0	1.5	4.0	203.9
Oct	5.1	4.9	3.3	1.0	0.2	0.3	14.9	71.0	66.2	52.5	14.0	1.5	4.1	209.4
Nov	5.4	5.3	4.9	1.3	0.1	0.3	17.3	62.7	66.4	52.3	17.6	1.5	4.1	204.7
Dec*	6.9	6.8	7.3	1.5	0.1	0.4	23.1	64.4	64.9	52.9	15.6	1.5	4.2	203.4
1991 Jan	5.7	5.5	6.3	1.4	0.2	0.3	19.5	65.3	71.0	55.3	15.5	1.4	4.1	212.6
Feb	5.9	5.8	6.9	1.5	0.1	0.3	20.6	60.8	69.9	53.3	18.0	1.3	4.1	207.4
Mar*	6.6	6.4	6.0	1.5	0.2	0.4	21.0	65.7	66.4	53.2	14.6	1.3	4.1	205.4
Apr[8]	4.9	5.4	4.3	1.1	0.2	0.3	16.2	64.4	72.0	50.5	13.9	1.3	4.1	206.3

1 Consumption by fuel producers *plus* disposals (including imports) to final users *plus* (for annual unadjusted figures only) net foreign trade and stock change in other solid fuels. See also footnotes 6 and 7 to Table 8.4.
2 Inland deliveries for energy use *plus* refinery fuel and losses *minus* the differences between deliveries to and actual consumption at power stations and gasworks.
3 Including non-energy use and excluding gas flared or re-injected.
4 Also temperature corrected.
5 Excludes generation from pumped storage stations.
6 Not seasonally adjusted.
7 For hydro the estimated annual out-turn.
8 Provisional.

Source: Department of Energy

From: Monthly Digest of Statistics, June 1991, Table 8.1

18.3 Coal supply and colliery manpower and productivity at BCC mines

	Thousand tonnes							BCC mines			
	Coal supply						Wage earners on colliery books (thousands)	Tonnes			
	Production							Average output[3] per manshift worked			
									Underground		
	Deep-mined	Opencast	Total[1]	Net imports	Import[2]	Export[2]		Overall	Total	Production[4]
	BHDC	BHDD	BHDB	BHDE	BHDF	BHDG	BHGA	BHGH	BHGI	BHGJ
1984	35 243	14 306	51 182	6 601	8 894	2 293	174	2.24	2.77	10.31
1985	75 289	15 569	94 112	10 300	12 732	2 432	148	2.49	3.12	11.36
1986	90 366	14 275	108 099	7 877	10 554	2 677	119	3.15	3.92	13.82
1987	85 957	15 786	104 533	7 428	9 781	2 353	102	3.59	4.42	15.81
1988	83 762	17 899	104 066	9 864	11 685	1 822	86	3.97	4.85	18.25
1989	79 628	18 657	101 135	10 088	12 137	2 049	66	4.33	5.21	20.41
1990[5]	71 501	17 802	92 937	12 290	14 783	2 493	59	4.52	5.39	21.86
1990 Apr	5 488	1 163	6 755	915	1 117	202	65	4.4	5.3	21.1
May	5 805	1 487	7 691	1 190	1 366	176	64	4.6	6.1	22.0
Jun*	6 976	1 913	9 271	1 195	1 416	221	63	4.6	5.6	22.3
Jul	5 586	1 402	7 480	557	691	133	62	4.7	5.6	22.5
Aug	4 671	1 434	6 358	1 198	1 374	175	62	4.4	5.2	21.6
Sep*	6 952	1 892	9 305	611	822	211	61	4.7	5.6	22.3
Oct	6 140	1 535	7 950	1 373	1 515	142	61	4.8	5.7	23.0
Nov	6 328	1 552	8 101	1 184	1 346	162	60	4.8	5.7	23.3
Dec*	5 781	1 451	7 318	1 669	1 855	186	59	4.4	5.2	22.1
1991 Jan	5 266	975	6 391	1 569	1 686	117	59	4.5	5.4	21.8
Feb	6 357	1 499	8 193	1 681	1 788	108	58	5.1	6.0	23.7
Mar*	7 999	1 902	10 196	1 705	1 906	201	57	5.2	6.2	25.0
Apr[5]	5 612	1 206	7 060	1 116	1 297	182	56	5.1	6.0	24.1

1 Including an estimate for slurry, etc, recovered and disposed of otherwise than by the British Coal Corporation (BCC).
2 As recorded in the *Overseas Trade Statistics*.
3 Saleable deep-mined revenue coal.
4 Output from production faces divided by production manshifts.
5 Provisional.

Source: Department of Energy

From: Monthly Digest of Statistics, June 1991, Table 8.4

ENERGY

18.4 Inland use and stocks of coal [1]

Stocks: end of period

Thousand tonnes

	Primary: collieries	Power stations[2]	Coke ovens	Other conversion industries[3]	Industry[4]	House coal[4,5]	Other[7]	Miscell-aneous[8]	Total inland consumption	Stocks[9]
		Fuel producers (consumption)			**Final users[6]**					
			Secondary			**Domestic**				
	BHEB	BHEC	BHED	BHEE	BHEF	BHEG	BHEH	BHEI	BHEA	BHEJ
1984	209	53 411	8 246	1 300	6 003	4 844	1 562	1 733	77 309	36 548
1985	332	73 940	11 122	2 176	7 474	6 533	2 102	1 707	105 386	34 979
1986	306	82 652	11 122	1 959	8 170	6 989	1 537	1 500	114 234	38 481
1987	235	86 177	10 859	2 052	7 986	5 685	1 475	1 425	115 894	33 157
1988	196	82 465	10 902	2 006	8 083	5 112	1 469	1 265	111 498	35 999
1989	146	80 633	10 792	1 717	7 514	4 344	1 368	1 066	107 581	39 083
1990[10]	115	82 555	10 822	1 526	7 510	3 876	1 240	1 191	108 834	37 553
1990 Apr	11	5 831	837	122	641	321	69	114	7 947	33 471
May	8	5 220	856	117	603	320	148	83	7 355	35 041
Jun*	9	6 297	1 058	146	729	349	123	104	8 815	36 718
Jul	8	4 849	872	105	539	300	103	64	6 839	38 139
Aug	6	5 212	822	110	493	224	81	59	7 007	38 648
Sep*	6	7 436	1 015	121	686	300	84	93	9 741	39 130
Oct	12	6 575	838	118	556	384	107	74	8 663	39 948
Nov	4	7 208	822	114	593	210	114	94	9 158	40 342
Dec*	15	9 403	1 001	119	675	347	106	127	11 792	37 553
1991 Jan	11	7 495	796	116	603	380	153	108	9 661	35 944
Feb	13	7 636	784	124	697	543	125	120	10 042	35 570
Mar*	13	8 708	970	158	779	412	144	118	11 302	36 101
Apr[10]	9	6 401	772	115	540	270	110	85	8 300	36 328

1 Stocks at end of period, Great Britain only.
2 Public supply and railway and transport power stations.
3 Low temperature carbonisation and patent fuel plants.
4 Includes estimated proportion of total imports.
5 Including miners' coal.

6 Disposals by collieries and opencast sites.
7 Anthracite, dry steam coal and imported naturally smokeless fuels.
8 Includes public administration and commerce.
9 Excluding distributed stocks held in merchants' yards, etc, mainly for the domestic market and stocks held by the industrial sector.
10 Provisional.

Source: Department of Energy

From: Monthly Digest of Statistics, June 1991, Table 8.5

18.5 Sources of supply and gas sent out by the gas supply system

Million therms

	Indigenous	Imported	Total into system[1]	Gas sent out
		Natural gas supply		
	Source			
	BHHB	BHHC	BHHA	BHHD
1985	14 314	5 003	19 317	19 047
1986	15 064	4 683	19 747	19 246
1987	15 783	4 416	20 198	19 814
1988	15 186	3 897	19 083	18 655
1989	14 976	3 852	18 828	18 590
1990[2]	16 492	2 724	19 215	19 175
1990 Mar*	1 977	407	2 384	2 372
Apr	1 524	151	1 674	1 674
May	872	139	1 011	972
Jun*	939	176	1 114	1 095
Jul	565	150	715	707
Aug	542	163	705	648
Sep*	1 006	192	1 197	1 173
Oct	1 163	148	1 311	1 286
Nov	1 695	231	1 925	1 931
Dec*	2 563	298	2 861	2 950
1991 Jan	2 140	235	2 375	2 548
Feb	2 308	244	2 552	2 794
Mar*	2 218	294	2 512	2 366
Apr[2]	1 582	236	1 818	1 709

1 Figures differ from Gas sent out because of stock changes and the inclusion of small quantities of Substitute Natural Gas and Town Gas in gas sent out. They include gas put to storage, but to avoid double counting, exclude gas withdrawn from storage to the system. The figures also differ from total consumption (expressed as oil equivalent in Table 8.1) because of producers own use and losses and small quantities not entering the gas supply system.
2 Provisional.

Source: Department of Energy

From: Monthly Digest of Statistics, June 1991, Table 8.6

18.6 Fuel used by and electricity production and availability from the electricity supply system

| | Million tonnes of oil or oil equivalent | | | | | Terawatt hours | | | | | | | |
| | Fuel used | | | | | | | Electricity supplied by type of plant | | | | | |
	Coal[2]	Oil[2,3]	Nuclear electricity	Hydro-electricity	Total[4]	Electricity generated	Own use[5]	Conventional Steam plant[6]	Nuclear	Other[7]	Total	Total Electricity available[8]
	FTAJ	FTAK	FTAL	FTAM	FTAN	BHJF	BHJJ	FTAB	FTAC	FTAD	BHJK	BHJL
1985	43.49	10.62	11.93	1.05	67.40	279.97	21.73	204.91	49.69	3.64	258.24	263.23
1986	48.62	6.08	11.40	1.20	67.31	282.26	21.10	209.98	47.48	3.70	261.16	270.88
1987	50.69	4.81	10.55	1.01	67.06	282.74	20.85	214.84	43.95	3.11	261.90	279.12
1988	48.51	5.39	12.41	1.21	67.53	288.51	21.58	211.50	51.70	3.73	266.93	285.16
1989	47.43	5.52	14.24	1.18	68.37	292.89	21.18	208.67	59.31	3.73	271.71	290.83
1990[9]	48.56	6.69	13.22	1.29	69.80	299.30	21.27	218.96	54.96	4.06	277.98	295.82
1990 May	..	..	..	..	..	20.39	1.36	14.45	4.41	0.15	19.03	20.43
Jun*	..	..	..	..	..	24.91	1.59	18.92	4.25	0.16	23.33	25.62
Jul	..	..	..	..	..	18.93	1.38	13.93	3.44	0.18	17.55	19.45
Aug	..	..	..	..	..	18.93	1.48	13.26	4.04	0.16	17.45	19.59
Sep*	..	..	..	..	..	25.69	1.86	18.51	5.05	0.27	23.83	25.97
Oct	..	..	..	..	..	22.50	1.61	16.43	4.05	0.41	20.89	22.60
Nov	..	..	..	..	..	25.40	1.82	18.22	5.04	0.34	23.58	25.35
Dec*	..	..	..	..	..	32.86	2.32	24.36	5.81	0.37	30.54	32.77
1991 Jan	4.41	0.37	1.34	0.15	6.27	27.36	1.96	19.35	5.57	0.48	25.40	27.18
Feb	4.49	0.66	1.38	0.06	6.59	28.97	2.06	20.98	5.75	0.18	26.91	28.83
Mar*	5.12	0.54	1.33	0.13	7.12	31.01	2.24	22.83	5.54	0.40	28.76	31.13
Apr[9]	3.77	0.41	0.99	0.13	5.30	23.21	1.60	17.08	4.12	0.43	21.62	23.49

1 Fuel used and electricity generated by major generating companies (National Power, PowerGen, Nuclear Electric, National Grid Company, Scottish Power, Scottish Hydro-Eleectric Scottish Nuclear, Northern Ireland Electricity service Midlands Electricity and South Western Electricity), and electricity available through the grid in England and Wales and from Distribution Companies in Scotland and Northern Ireland.
2 Including quantities used in the production of steam for sale.
3 Including oil used in gas turbine and diesel plant and for lighting up coal-fired boilers and Orimulsion.
4 Including wind power and refuse derived fuel.
5 Used in works and for pumping at pumped storage stations.
6 Coal Oil (including Orimulsion) and mixed or dual-fired (including gas).
7 Including gas turbine, diesel, wind and hydro-electric plant.
8 Including net imports and purchases from outside sources mainly UKAEA and British Nuclear Fuels plc, and net of supplies direct from generators to final consumers.
9 Provisional.

Source: Department of Energy

From: Monthly Digest of Statistics, June 1991, Table 8.7

18.7 Deliveries of petroleum products for inland consumption
Thousand tonnes

| | | Naphtha (LDF) and Middle Distillate Feedstock[2] | Motor Spirit | | Kerosene | | | Gas/diesel oil | | | | | |
| | | | | | Aviation turbine fuel | Burning oil | | | | | | | |
	Butane and propane[1]		Total	of which: Unleaded		Premier	Standard domestic	Derv fuel	Other	Fuel oil	Lubricating oils	Bitumen	Total[3]
	BHOB	BHOC	BHOD	BHON	BHOE	BHOF	BHOG	BHOI	BHOJ	BHOK	BHOL	BHOM	BHOA
1985	1 543	3 792	20 403	..	5 007	126	1 277	7 106	9 733	15 979	816	1 887	69 781
1986	1 885	3 786	21 470	..	5 497	114	1 418	7 866	9 241	12 514	803	2 019	69 227
1987	1 838	3 640	22 184	..	5 815	100	1 390	8 469	8 608	9 935	828	2 162	67 701
1988	1 913	3 866	23 249	258	6 200	68	1 415	9 370	8 456	11 865	849	2 342	72 317
1989	1 893	3 932	23 928	4 652	6 564	55	1 417	10 118	8 323	11 125	878	2 423	73 035
1990[4]	1 969	3 477	24 312	8 255	6 589	41	1 526	10 652	8 046	11 997	822	2 491	73 943
1990 Feb	190	346	1 857	559	453	5	162	846	754	926	62	165	5 976
Mar	132	350	2 144	668	508	6	156	982	810	1 469	77	231	7 079
Apr	154	245	2 001	648	546	5	138	842	682	1 034	64	183	6 062
May	104	227	2 140	704	579	1	80	912	585	1 235	72	238	6 336
Jun	121	270	2 026	681	612	1	79	904	541	1 288	69	253	6 330
Jul	136	301	2 112	716	654	1	98	918	631	1 229	65	232	6 545
Aug	133	310	2 179	762	653	1	139	930	623	837	75	234	6 264
Sep	138	279	1 920	700	589	3	95	850	591	691	75	199	5 571
Oct	158	190	2 041	767	563	2	85	906	626	629	79	262	5 663
Nov	174	273	2 023	761	467	4	134	938	703	740	64	219	5 881
Dec	153	326	1 951	725	475	6	199	807	733	667	55	129	5 651
1991 Jan	148	437	1 933	736	446	8	253	924	926	849	73	134	6 312
Feb	155	330	1 686	651	394	8	186	780	817	1 214	53	125	5 907
Mar	100	360	2 051	802	418	3	128	891	653	1 001	65	216	6 090
Apr[4]	206	293	2 011	809	430	2	145	907	691	871	66	193	6 022

1 Including amounts for petro-chemicals.
2 Now mainly petro-chemical feedstock. Prior to the October 1986 issue of the *Monthly Digest*, Middle Distillate Feedstock was included in the Gas/Diesel (Other) column.

3 Including other petroleum gases, aviation spirit, wide-cut gasoline, industrial and white spirits, petroleum wax, non-domestic standard burning oil and miscellaneous products, but excluding refinery fuel.
4 Provisional.

Source: Department of Energy

From: Monthly Digest of Statistics, June 1991, Table 8.10

Definitions and sources

Industry grouping: The industries named in the first column of the table represent the classes of the Standard Industrial classification, as revised in 1980.

Employment: UK Employment figures (except classes 21 and 23) are averages of the four end-quarter months' figures for GB (Department of Employment) and NI (Office of Manpower Economics). Figures for classes 21 and 23 are based on the Census of Employment 1987.

Index of Production: The official measure of value added by each industry, at constant 1985 prices; the figures therefore represent a measure of volume.

Exports and Imports: The series are derived from *Overseas Trade Statistics in the United Kingdom* (OTS) where every tariff code is allocated to an industry

Import Penetration Ratio: Imports as a percentage of UK Demand (Production plus imports minus exports).

For other sources see:

Guide to Official Statistics, 1990 edition (200 pages approximately fully indexed) HMSO

19.1 Indicators of UK Manufacturing Industrial Activity

SIC (80) Class		UK Employees (thousands)	Index of Production (1985=100)	Exports f.o.b. (£m)	Imports c.i.f. (£m)	Import Penetration (%)
21	1985	. .	100.0	36	907	100
Extraction of	1986	. .	88.0	73	716	. .
Metalliferous	1987	. .	80.4	17	749	. .
ores	1988	. .	76.1	23	601	. .
	1989	. .	70.4	28	700	. .
	1990	. .	67.5	22	816	. .
22	1985	180	100.0	3860	4065	39
Metal	1986	158	101.5	3953	4413	43
Manufacturing	1987	145	109.3	4097	4278	24
	1988	136	122.7	4325	5372	29
	1989	134	125.5	5254	6426	. .
	1990	158	122.5	5589	6145	. .
23	1985	. .	100.0	1209	1544	43
Extraction of	1986	. .	101.0	1554	1789	37
minerals n.e.s.	1987	. .	103.4	1648	1674	36
	1988	. .	109.4	1924	2045	33
	1989	. .	107.8	2041	2083	. .
	1990	. .	99.7	1945	2107	. .
24	1985	217	100.0	1120	891	13
Non-Metallic	1986	203	102.2	1127	989	13
mineral	1987	194	107.9	1183	1140	16
products	1988	195	119.0	1254	1377	17
	1989	203	121.3	1377	1587	. .
	1990	205	115.3	1510	1602	. .
25	1985	332	100.0	9471	7034	40
Chemicals	1986	323	101.5	9617	7311	40
	1987	317	108.8	10539	8358	41
	1988	322	114.4	11331	9267	41
	1989	327	119.3	12522	10478	. .
	1990	325	118.2	13433	10935	. .
26	1985	10	100.0	435	383	71
Man-Made fibres	1986	8	103.6	416	420	68
	1987	7	109.9	424	473	66
	1988	7	107.2	464	469	67
	1989	7	114.6	566	491	. .
	1990	7	118.1	656	497	. .

Table 19.1 (continued)

SIC (80) Class		UK Employees (thousands)	Index of Production (1985 = 100)	Exports f.o.b. (£m)	Imports c.i.f. (£m)	Import Penetration (%)
31	1985	329	100.0	1218	1380	16
Metal goods	1986	324	99.5	1180	1519	16
n.e.s.	1987	325	103.3	1287	1778	18
	1988	335	111.7	1439	1992	19
	1989	334	112.8	1635	2277	..
	1990	321	110.7	1909	2414	..
32	1985	766	100.0	8295	6158	36
Mechanical	1986	749	96.5	8385	6593	37
Engineering	1987	746	96.8	8984	7578	38
	1988	772	105.6	9281	9023	40
	1989	774	110.3	10286	10264	..
	1990	756	111.8	11838	10262	..
33	1985	87	..	3543	4213	100
Office	1986	83	..	3344	4230	100
machinery &	1987	83	..	4211	5098	93
data processing	1988	84	..	4929	5828	91
equipment	1989	83	..	5716	7041	..
	1990	81	..	6027	7246	..
34	1985	596	100.0	7025	8246	47
Electrical &	1986	574	101.7	7414	8732	47
electronic	1987	567	103.1	7993	10018	49
machinery	1988	563	111.2	8995	11461	48
	1989	567	117.7	10370	13157	..
	1990	566	117.5	11809	13328	..
35	1985	274	100.0	4066	6963	50
Motor vehicles	1986	264	96.8	4076	8178	51
and parts	1987	262	104.2	5003	9058	48
	1988	272	119.5	5533	11692	48
	1989	269	125.3	6754	13488	..
	1990	249	121.1	8045	12972	..
36	1985	288	100.0	4178	3035	45
Other transport	1986	275	112.0	4642	2889	45
equipment	1987	257	113.1	5478	3212	42
	1988	241	108.4	5832	4427	55
	1989	243	128.9	7556	4554	..
	1990	258	129.5	7872	6373	..
37	1985	105	..	1194	1503	57
Instrument	1986	105	..	1230	1617	56
engineering	1987	103	..	1346	1696	58
	1988	102	..	1551	1954	58
	1989	97	..	1739	2222	..
	1990	92	..	1703	2186	..
41/2	1985	594	100.0	4058	7294	18
Food, drink	1986	577	100.8	4161	7666	18
tobacco	1987	571	103.2	4741	7951	18
	1988	564	105.0	4697	8172	15
	1989	552	105.2	5403	8981	..
	1990	543	106.1	5945	9644	..
43	1985	247	100.0	1859	3192	44
Textiles	1986	248	100.4	1855	3400	45
	1987	240	104.9	2069	3831	47
	1988	237	102.1	2132	4108	48
	1989	217	97.1	2315	4302	..
	1990	201	92.3	2483	4516	..
44	1985	22	100.0	260	406	49
Leather and	1986	21	107.0	273	420	46
leather goods	1987	21	109.6	345	499	49
	1988	21	101.5	322	502	49
	1989	20	100.5	374	553	..
	1990	20	87.7	394	578	..

INDUSTRY

Table 19.1 (continued)

SIC (80) Class		UK Employees (thousands)	Index of Production (1985 = 100)	Exports f.o.b. (£m)	Imports c.i.f. (£m)	Import Penetration (%)
45	1985	314	100.0	1165	2488	35
Footwear and	1986	312	101.2	1210	2715	36
clothing	1987	311	101.5	1405	3114	39
	1988	314	102.4	1344	3379	39
	1989	304	100.1	1384	3760	. .
	1990	304	100.3	1633	4223	. .
46	1985	215	100.0	397	2124	30
Timber and	1986	222	103.4	380	2383	31
wooden	1987	232	112.5	416	2751	31
furniture	1988	243	126.7	395	3100	30
	1989	250	126.7	443	3247	. .
	1990	253	123.1	520	3211	. .
47	1985	486	100.0	1620	3537	21
Paper, printing	1986	478	104.3	1704	3768	21
and publishing	1987	480	114.4	2004	4566	22
	1988	488	124.9	2097	5138	21
	1989	494	131.5	2359	5780	. .
	1990	490	133.9	2847	5727	. .
48	1985	196	100.0	1522	1895	26
Rubber and	1986	199	107.0	1646	2180	27
plastics	1987	205	119.9	1947	2652	28
processing	1988	216	132.2	2009	2810	26
	1989	220	139.1	2266	3118	. .
	1990	223	142.4	2603	3319	. .
49	1985	77	100.0	1051	1300	38
Other	1986	76	101.0	1195	1589	39
manufacturing	1987	75	104.4	1303	1646	46
industries	1988	74	115.7	1205	1909	44
	1989	78	124.3	1388	2403	. .
	1990	80	128.7	1448	2402	. .

Source: Industrial Economic Indicators Database (DTI)

Key Data 91, © Crown copyright 1991

INDEX

INDEX

Key Data 91, © Crown copyright 1991

GOVERNMENT STATISTICS
A BRIEF GUIDE TO SOURCES

INTRODUCING THE GOVERNMENT STATISTICAL SERVICE

The Government Statistical Service (GSS) comprises the statistics divisions of all major departments plus the two big collecting agencies, the Central Statistical Office, and the Office of Population Censuses and Surveys[1]. The Central Statistical Office (CSO) co-ordinates the system. In sum, it is the greatest concentration of statistical expertise and by far the largest single provider of statistics in the country.

The GSS exists to service the needs of government. But much of the information it compiles is readily usable in business management, particularly in marketing. The statistics are not restricted to the economic field. There is also a considerable body of social statistics which can be used by businesses, as well as local authorities, universities and others concerned with current problems and changes in society. A vast amount of this information is published regularly and more is available on request.

The price of each publication covers the cost of making the information available. The total cost of preparing the statistics is much greater; in fact, you or your firm may already have contributed to this by co-operating in various statistical enquiries. So it is in everyone's interest to see these statistics used as widely as possible, subject always to safeguarding the confidentiality of individual data.

What the GSS normally *cannot* do is give tailor-made answers to an individual firm's problems. It simply provides a very large source of information which may help in finding solutions; and it is nearly always worth checking to see what is available before embarking on private research. This booklet is intended to help would-be users do just that.

USING GOVERNMENT STATISTICS

Many firms make use of government statistics together with information from other sources to help both in broad strategic planning and in day-to-day decision-making. The following checklist indicates some of the main areas of use which might be applicable to your firm's needs.

Marketing
- Look at your industry's *Business Monitor, CSO Bulletin, Ministry of Agriculture Information Notice* or *Housing and Construction Statistics* to compare your own performance against general sales trends and to watch for opportunities for diversification.

1 The OPCS also includes the registration division of the Registrar General's Office.

- From a variety of sources you can check the number of potential customers in a given sector and compare characteristics with your own customers to highlight possible weaknesses in marketing strategy.
- If you are in consumer lines, trends in expenditure are available from the *National Food Survey* and the *Family Expenditure Survey.*
- For test marketing, the *Census of Population* can provide very localised data on numbers of consumers.
- Some statistics are available on a regional or area basis, for example *Regional Trends,* and can assist in determining quotas for area salesmen.
- Watch the foreign competition by looking at the import figures available from agents appointed by HM Customs and Excise to market the overseas trade statistics.

Contracts

Many contracts contain cost escalation clauses. Consult the Property Services Agency's *Price Adjustment Formulae for Construction Contracts, Monthly Bulletin of Construction Indices,* or speak to the Central Statistical Office and Department of Employment respectively to find the most appropriate materials and labour indices to use.

Accounts

Price Index Numbers for Current Cost Accounting provides a historical series of price index numbers for use in this system of company accounting, together with an explanation of the purpose, content and method of compiling the indices. Latest index numbers are published each month in *Business Monitor MM17 (Price Index Numbers for Current Costing Accounting) (Monthly Supplement).*

Buying

Watch the sales trends of your products in *Business Monitors*, and their prices with the appropriate producer price indices from the Central Statistical Office.

Personnel

The *Employment Gazette* and other Department of Employment publications will keep you fully informed on the latest figures and trends in earnings, labour costs, overtime, employment, unemployment, vacancies, hours of work, stoppages due to industrial disputes, retail prices, tourism, etc. — in some cases analysed by industry and region.

Management efficiency and finance

You can compare your own costs and operating ratios with those for your industry as a whole, for example:
- net output per head, stocks as a percentage of sales, wages per £ of total sales, etc., from the annual *Census of Production* reports;

- return on capital, dividends and interest as a percentage of assets, profits as a percentage of turnover etc., from *Business Monitor MA3 (Company Finance)* and *MO3 (Finance of Large Companies).*

Investment

Major investment decisions are made against the background of trends and prospects in the economy. Regular press notices on output, demand, earnings, prices, unemployment, trade, etc., are released by government departments. The data they contain are all brought together once a month in *Economic Trends.*

Social change

Firms should also be aware of major changes that are taking place in society which may in time affect the market for their goods and environment in which they operate. Each year, *Social Trends* records trends and distributions in all areas of social concern — population, households, education, health, housing, environment and leisure are just a few examples. *OPCS Monitors* and reports on the *General Household Survey* also provide valuable information in this field.

HOW TO FIND WHAT YOU WANT

For anyone coming new to government statistics, the first thing to understand is that each department prepares and publishes its own statistics — via HMSO in the case of most printed publications.

If the series is of sufficient interest, it may emerge first in summary form as a press notice. A more detailed treatment is given in the department's own publications, such as *Business Monitors* and *CSO Bulletins,* which are purely statistical reports. Finally, it may be included in all-embracing statistical periodicals like the *Monthly Digest of Statistics* or the annual *Social Trends,* again often in a more summarised form.

It may be that your needs can be met from one of these sources; or you may have to consult the special departmental publication; or you may have to go further still and seek a special tabulation — at a reasonable cost — from the department. If the published source does not yield what you want it is usually worth checking with the department to see if they can help you.

The following pages list the main publications. Current copies of selected series can be referred to at many public reference libraries, and at the Central Statistical Office Library, Newport, Gwent. The Library at OPCS holds a wealth of national and international data on social subjects. Most of the principal volumes will also be held or quickly obtained by major public libraries around the country. If you wish to purchase a copy, this can be done through any of the HMSO Bookshops listed on the inside back cover and through booksellers. Individual copies of monthly and quarterly and PAS series *Business Monitor* titles can be purchased only from the Central Statistical Office Library (see below). Subscriptions are handled by HMSO. If you cannot find what you need, refer to the more comprehensive *Guide to Official Statistics* or contact the appropriate departmental source, listed on pages 105 to 110. All the extension numbers were up-to-date at the time of going to print. However, they are subject to change, and if you have difficulty in finding the right person you should phone or write to:
The Library, Central Statistical Office, Cardiff Road, Newport, Gwent NP9 1XG (Telephone 0633 812973; fax 0633 812599) or
Press, Publications and Publicity Office, Central Statistical Office, Great George Street, London SW1P 3AQ (Telephone 071-270 6363/6364).

IN SUMMARY

- **Look at the published source**
  at your local reference library
  or the Central Statistical Office Library
  or by purchasing from HMSO or through booksellers.

- **For further information**
  phone or write to the appropriate department.

- **If in doubt**
  consult the *Guide to Official Statistics*
  phone or write to the Central Statistical Office.

GOVERNMENT STATISTICAL PUBLICATIONS

The most important government publications containing statistics are listed on pages 93 to 104. In most cases they are statistical publications like *Business Monitors;* in others, the statistics appear in addition to other information. Far more detail, including information about *ad hoc* reports and articles, is offered by the *Guide to Official Statistics,* 5th Edition, ISBN 011 620394 3 (HMSO, £24.00), which can be seen in many major libraries.

Some statistics are compiled separately for England and Wales, Scotland and Northern Ireland, and it is not always possible to combine them into United Kingdom totals. In these cases it is often necessary to consult the separate publications for detailed information about each.

Unless otherwise stated, all publications listed are published by HMSO and can be ordered from any of the bookshops listed on the inside back cover. It may be worth checking whether bodies such as trade associations or the National Economic Development Office (NEDO) have published any recent reports on the industry in question. All NEDO publications are listed in the free booklet *NEDO books catalogue,* obtainable from NEDO Books, Millbank Tower, Millbank, London SW1P 4QX, telephone 071-217 4037/4036.

Other bodies publishing statistics in special subject areas include: the Bank of England, for certain economic and financial statistics, in its *Quarterly Bulletin,* published by the Economics Division, Bank of England, London EC2R 8AH, telephone 071-601 4030; and the Statistical Information Service of the Chartered Institute of Public Finance and Accountancy, 3 Robert Street, London WC2N 6BH, telephone 071-930 3456, for a wide range of local government statistics.

Note:
All prices were up-to-date at the time of going to press, but are subject to change and should be checked.

General background and reference

Publications which, while not themselves sources of statistics, provide useful background, or are helpful or even essential references for some users of statistics.

Guide to Official Statistics, No 5, Revised 1990
(mentioned on page 92)
£24.00

Statistical News
Provides a comprehensive account of current developments in British official statistics. Shorter notes give news of the latest developments in many fields, including international statistics. A cumulative index in each Winter issue provides a permanent and comprehensive guide to development in all areas of official statistics.
Quarterly £4.75. Annual subscription (including postage) £17.50

United Kingdom National Accounts
Sources and Methods
An essential reference book for everyone who makes use of the national accounts data published in *United Kingdom National Accounts* — the 'CSO Blue Book' — or in *Economic Trends*. It contains details of the concepts, definitions, statistical sources, methods of compilation and reliability of the various statistical series which comprise the national accounts.
£14.95

Standard Industrial Classification Revised 1980
The basis of classification for all industries and activities used in the National Accounts, the Annual Census of Production and many other statistical publications.
£4.25

Indexes to the Standard Industrial Classification
Revised 1980
£12.95

A Guide to Northern Ireland Statistics [1]
Occasional Paper 16
£5.00

Scottish Office statistical information leaflet [2]
Annual, free

General digests

Monthly Digest of Statistics
Collection of main series from all government departments.
Monthly £6.50. Annual subscription (including Supplement and postage) £72.00

Annual Abstract of Statistics 1991
Contains many more series than the *Monthly Digest* and provides a longer run of years.
Annual £19.95

Social Trends 1991
Brings together key social and demographic series in colour charts and tables.
Annual £23.50

Regional Trends 1991
A selection of the main statistics that are available on a regional basis.
Annual £23.00

Scottish Abstract of Statistics No. 18 1989 [3]
Annual £16.00

Scottish Statistics fact card
Annual, free

Digest of Welsh Statistics No. 36 1990 [4]
Annual £6.00

Northern Ireland Annual Abstract of Statistics [1]
Annual. Price on application

Welsh Social Trends No. 8 1991 [4]
Biennial £9.50

The economy

(See also Transport etc. page 97 and Manpower etc. page 99.)

General economic

Economic Trends
Commentary and selection of tables and charts provide a broad background to trends in the UK economy. The Annual Supplement presents very long runs of quarterly figures for all the main series.
Monthly £11.00. Annual subscription (including Supplement and postage) £130.00

---

1   Obtainable from Policy Planning and Research Unit, Room 250, Parliament Buildings, Stormont, Belfast BT4 3SW

2   Obtainable from Central Statistical Unit, Room 5/52, New St. Andrew's House, Edinburgh EH1 3SX

3   Obtainable from Scottish Office Library (address on page 109)

4   Obtainable from E & SS Division, Welsh Office, Cathays Park, Cardiff CF1 3 NQ.

Scottish Economic Bulletin
Contains an economic review, articles on the Scottish economy and regular statistics.
6-monthly £9.50

Statistical Bulletins [2] on various aspects of the Scottish economy. Issued occasionally, covering the index of production and construction, overseas-owned firms, the electronics industry, offshore employment in the northern North Sea, and other topics. The *Scottish Economic Bulletin* lists recent Statistical Bulletins (eg 'Offshore employment in the northern North Sea in 1989', 'The electronics industry in Scotland').
Occasional £1.25.

Welsh Economic Trends No. 12 1990 [1]
Biennial £9.50

CSO Bulletins
The Central Statistical Office publishes a series of *CSO Bulletins*, which contain data previously published as articles in *British Business*; details and order forms are available from the Librarian, Central Statistical Office, Cardiff Road, Newport, Gwent NP9 1XG. Telephone 0633 812828. Fax 0633 812599.

UK National Accounts ('CSO Blue Book') 1991
The essential data source for everyone concerned with macro-economic policies and studies. The principal annual publication for national accounts statistics providing detailed estimates of national product, income and expenditure for the United Kingdom. The Blue Book covers value-added by industry, personal sector, public corporations, central and local government, capital formation and financial accounts.
Annual £13.95

UK Balance of Payments ('CSO Pink Book') 1991
The basic reference book for balance of payments statistics, with information on visible trade, invisibles, investment and other capital transactions and official financing as well as sections on specific aspects of the balance of payments, detailed balance of payments data for the last 11 years, summary figures for earlier years and full notes and definitions.
Annual £11.75

Financial Statement and Budget Report
Published by the Treasury on Budget day, it sets out the Government's economic and financial strategy and reviews the performance and prospects of the economy. It describes the Budget tax measures and estimates the income and expenditure of the public sector in the current and the forthcoming year.
Annual, price on application

The Autumn Statement
Published by the Treasury each year, it announces the Government's spending plans for the next three years, sets out the prospects for the economy, proposals resulting from the annual review of National Insurance Contributions, and a tax ready reckoner showing various illustrative tax changes on tax receipts.

Input-Output Tables for the UK, 1984
An important reference book for the understanding and analysis of economic inter-dependence of the various industrial sectors. It presents a snapshot of the economy for the year 1984, the first such picture since the last set of tables for 1979.
May 1988 £19.95 including IBM PC compatible 5¼" floppy disk.

Scottish Input-Output Tables for 1979 [2]
Vol 1: free; vols 2-5 £5 each

Economic Progress Report [3]
Published by the Treasury. It contains articles on economic subjects, analysing or giving background to government economic policy. From time to time it also contains supplements on subjects of special interest, eg the Budget.
Bi-monthly, free

CSO Macro-Economic Databank
Data stored cover the index of output of production industries, national accounts, balance of payments, prices, earnings, employment, financial statistics and cyclical indicators. Data are available on magnetic tape or in computer print-out form. Contact: 071-270 6386.

Financial and companies
(See also Public sector)

Financial Statistics
Key financial and monetary statistics of the UK.
Monthly £8.50. Annual subscription (including Handbook and postage) £100.00

Occupational Pension Schemes, 1983
£5.00

Price Index Numbers for Current Cost Accounting (Monthly Supplement) (Business Monitor MM17)
On subscription £32 .50 per annum including 1st class postage. Also available on magnetic tape.

Price Index Numbers for Current Cost Accounting Summary Volume 1983–1987 (Business Monitor MO18)
£8.90

Aerospace and Electronics Costs Indices (1980=100 and 1985=100) (Business Monitor MM19)
Includes price indices for materials and fuel, earnings and combined costs.
Monthly. £29.50 per annum including postage

Size Analyses of United Kingdom Businesses (Business Monitor PA 1003)
Analysis of company size by sector, by turnover size for service industries and turnover and employment size for manufacturing industry.
Annual, 1990 edition £15.00

Engineering Sales and Orders [4]
Monthly. £30 per annum including postage

Company Finance (Business Monitor MA3)
An analysis of the accounts of large and small companies.
Annual, 21st issue £10.00

---

1 Obtainable from E & SS Division, Welsh Office, Cathays Park, Cardiff CF1 3NQ.
2 Obtainable from Scottish Office Library (address on page 109).
3 Obtainable from Economic Progress Report (distribution), COI, Hercules Road, London SE1 7DU.
4 Obtainable from the Central Statistical Office Room 1-372, Cardiff Road, Newport, Gwent NP9 1XG. Telephone 0633 812873.

Finance of Large Companies
(Business Monitor MO3)
Analysis of the 500 largest companies.
Twice-yearly. Available in 1991 as a package with MA3. Price on application

Insurance Companies' and Private Pension Funds' Investments (Business Monitor MQ5)
Quarterly. £18.00 per annum including postage

Credit Business (Business Monitor SDM6)
Data on consumer and business credit, including new credit granted, and amounts outstanding. Analysis by type of grantor.
Monthly. £29.50 per annum including postage

Assets and liabilities of finance houses and other consumer credit grantors (Business Monitor SDQ7)
Data on holdings & acquisitions by type of asset.
Quarterly. £18.00 per annum including postage

Public sector
(See also **General economic**, page 93, particularly *Financial Statement and Budget Report* and *UK National Accounts* and Taxation, page 100.)

Departmental Reports
A series of 19 individual reports (published by spending Departments) plus a statistical supplement, which together present a detailed and comprehensive picture of the Government's expenditure plans as announced by the Chancellor of the Exchequer in his Autumn Statement.

Supply Estimates
Gives details of estimated cash requirements of central government departments for the forthcoming financial year.
Annual, price on application

Annual Review of Government Funded R&D
Presents government expenditure on R&D, both outturn and forward plans by department, R&D performed by industry, total UK R&D and international comparisons of R&D.
Annual £10.95

Summary analyses in:
Supply Estimates — Summary and Guide
Annual, price on application

Appropriation Accounts (12 volumes)
Give a comparison between the actual expenditure of government departments during the previous financial year and the provision in the Supply Estimates for that year, together with the report of the Comptroller and Auditor General upon them. (See also *UK National Accounts*, page 94 and *Financial Statistics*, page 94.)
Annual, price on application

Cash Limits 1986-87 Provisional Outturn White Paper
Contains provisional outturn figures for government expenditure subject to cash limits and running costs limits in 1986-87; and final outturn figures for cash limited expenditure in 1985-86.
Annual £6.30

Local Government Financial Statistics, England, No. 2 1990
Annual £8.00

Scottish Local Government Financial Statistics [1]
Annual £4.00 (Historical time series now available)
£6.00

Welsh Local Government Financial Statistics No. 14 1990 [2]
Annual £6.00

District Councils, Summary of Statements of Accounts, Northern Ireland
Annual, price on application

Production industries

Report on the Census of Production
Published annually in the Business Monitor PA series in separate parts for each industry, with a separate volume of summary tables. They include data on total purchases, total sales, stocks, work in progress, capital expenditure, employment, and wages and salaries.
Each part individually priced

Purchases Inquiry (Business Monitor PO1008)
Materials and fuels purchased by manufacturing and energy industries in 1984.
£17.50

Statistics of Manufacturers' Sales
Published in the Business Monitor series, mainly on an annual basis (PAS series), with some industries covered quarterly and monthly (PQ and PM series).
Subscriptions to individual industries £7.50 per annum (PAS series), £18.50 per annum (PQ series) £29.50 per annum (PM series). Complete set £825 per annum.

Manufacturing industry, import penetration and export sales ratios (Business Monitor MQ12)
Gives annual moving averages of these ratios over nine-quarter spread for manufacturing industries.
Quarterly. Annual subscription £18.00 including postage

UK Directory of Manufacturing Businesses, 1989 (Business Monitor PO1007)
Compiled from the register of manufacturing businesses in the UK which is held by the Central Statistical Office for conducting statistical enquiries. Details from the Library, Central Statistical Office, Cardiff Road, Newport, Gwent NP9 1XG.
Issued in 6 volumes, £15.30 each. Also available on magnetic tape

UK Directory of Manufacturing Businesses, 1990 Supplement (Business Monitor PO 1007A)
Additional entries to supplement Business Monitor PO 1007.
£26.90

Census of Production for Northern Ireland [3]
Annual, price on application

Digest of UK Energy Statistics
Contains tables and charts of UK energy production and consumption. Separate sections deal with production and consumption of individual fuels, oil and gas reserves, fuel prices and foreign trade in fuels. Note: Further statistical information can be obtained from the annual report and accounts of each of the nationalised fuel industries and British Gas plc.
Annual, 1990 Edition £15.75

---

1 Obtainable from Scottish Office Library (address on page 109).

2 Obtainable from E & SS Division, Welsh Office, Cathays Park, Cardiff CF1 3NQ.

3 From HMSO Bookshop, 80 Chichester Street, Belfast BT1 4JY.

Energy Trends [1]
Contains monthly and quarterly tables of production and consumption of fuels, aggregated energy consumption and fuel prices together with a commentary.
Monthly, on application

Minerals (Business Monitor PA 1007)
Presents statistics on minerals extracted, employment etc., in the mining and quarrying industry.
Annual, 1988 edition £8.20

Survey of Scientific Research and Development carried out within the UK (Business Monitor MO14)
R&D expenditure by industry sector.
1985 edition £6.00

United Kingdom Minerals Yearbook 1990 [2]
Presents commentary for 1990 and comprehensive production, consumption and trade statistics on UK minerals from 1985 to 1989.
Annual £25.00

World Mineral Statistics, 1984-88 [2]
Presents world production and trade statistics on 65 mineral commodities from 1984 to 1988.
Annual £40.00 including postage by surface mail

(Survey of land for mineral working, see page 104; for UK external trade in minerals, see page 97.)

Housing, construction and property

Construction (Business Monitor PA 500)
Contains data on output and costs, capital expenditure, employment and various operating ratios.
Annual, 1988 edition £7.85

Housing and Construction Statistics, 1979-1989, Great Britain
Annual £21.00

Housing and Construction Statistics, Great Britain
Quarterly in 2 parts £4.10 each.
Annual subscription £31.50

Local Housing Statistics, England and Wales
Quarterly, individually priced

English House Condition Survey 1986
Report of the Physical Condition Survey
£20.00

Housebuilding in England by Local Authority Areas; 1980-1988 [7]
Annual £6.50

Statistical Bulletins on Housing, Scotland [3]
'Housing trends in Scotland', issued quarterly.
Also *ad hoc* topic Bulletins, eg household estimates, public sector sales.
£1.25

Northern Ireland Housing Statistics
Annual, price on application

Welsh Housing Statistics No. 10 1990 [4]
Annual £4.00

Welsh House Condition Survey 1986 [4]
Periodic £4.00

Monthly or quarterly releases are also available[5] on housebuilding, housing renovations, construction new orders, building materials and components, brick production, new orders by type of work, construction indices for use with price adjustment formulae, and civil engineering work price adjustments for building contracts.

Commercial and Industrial Floorspace Statistics, Wales No 10 1989 [4]
Annual £5.00

Local authorities' action under the homelessness provisions of the 1985 Housing Act: England [6]
Quarterly

Agriculture and Fisheries

Ministry of Agriculture, Fisheries and Food publishes much of its information through *Statistical Information Notices and in a more permanent form in:*

Agriculture in the United Kingdom 1989
Annual £10.30

Agricultural Statistics: United Kingdom, 1988
Annual £12.00

Farm Incomes in the United Kingdom, 1990
Annual £16.70

Household Food Consumption and Expenditure; Report of the National Food Survey Committee 1989
Annual £19.75

Agricultural Market Report (Livestock and Horticultural produce)
Weekly, price on application

Basic Horticultural Statistics for the United Kingdom 1979-1988
Annual, free
see also
Welsh Agricultural Statistics No 12 1990 [4]
Annual £5.00

Farm Incomes in Wales No 4 1990
Annual £5.00

Agriculture in Scotland 1989 [3]
Annual £6.55

Economic Report on Scottish Agriculture 1989 [3]
Annual £4.00

---

1 From Department of Energy, EcS, Room 3.3.26, (address on page 107).
2 Obtainable from British Geological Survey, Keyworth, Nottingham NG12 5GG.
3 From Scottish Office Library (address on page 109).
4 Obtainable from E & SS Division, Welsh Office, Cathays Park, Cardiff CF1 3NQ.
5 Further details from Department of the Environment, LGS4 Division, 2 Marsham Street, London SW1P 3EB. Telephone: 071-276 4003.
6 From Department of the Environment, HDS3 Division, Room N2/12, 2 Marsham Street, London SW1P 3EB.
7 From the Department of the Environment Publication Sales Unit, Victoria Road, Ruislip HA4 0AZ. Telephone: 081-841 3425.

Statistical Review of Northern Ireland Agriculture
1988 [1]
Annual £10.00

DAFS Statistical Bulletin No 1
(a) 1990 - Chernobyl Accident; Monitoring for
Radioactivity in Scotland [2]
£1.50

English and Welsh Fishing 1988
Annual free

UK Sea Fisheries Statistical Tables 1988
Annual £10.00

Scottish Sea Fisheries Statistical Tables, 1988 [3]
Annual £6.00

Scottish Fishing Fleet, 1989 [3]
Annual £6.00

Scottish Inshore Fishing Fleet — Cost and Earnings
Survey 1977-1983 [2]
Occasional, £2.00

DAFS Statistical Bulletin — Trends in Fish Landings
1960-1982 [3]
Occasional, free

DAFS Statistical Bulletin — Scottish Salmon and Sea
Trout Catches 1989 [3]
Annual £1.25

Defence

Defence Statistics (Statement on the Defence
Estimates, Volume II)
Separate sections cover the structure of the Armed
Forces, defence finance and trade in defence equipment,
the equipment programme, manpower, welfare and
defence services provided.
Annual £7.00

External trade

(See also *UK Balance of Payments* page 94, and *Digest
of UK Energy Statistics* page 95; for export sales ratios,
etc see *Manufacturing industry, import penetration and
export sales ratios* page 95; for mineral statistics see
page 96.)

Overseas Trade Statistics of the United Kingdom
Business Monitor MM20/MA20
Gives detailed statistics of exports and imports
cumulated throughout the year. Annual volumes are also
produced.
*Monthly (MM20) January to November £30.00;
December edition £43.00. Annual Revised Edition
(MA20) £43.00. Annual subscription £335*

Overseas Trade Analysed in Terms of Industries
(Business Monitor MQ10)
Gives an analysis of commodities imported and exported,
according to the industries of which they are principal
products.
Quarterly. Annual subscription £18.00 inc. postage
With a few exceptions, analyses of exports and imports
under individual trade headings and with individual
countries or through United Kingdom ports for particular
months or cumulative periods, can be obtained from
Marketing Agents appointed by HM Customs and Excise.
A list of names and addresses is available from the
address on page 106. Certain of these Marketing Agents
also publish importers' names and addresses against
appropriate 9 digit Commodity Codes of the UK Tariff.

Overseas Transactions (Business Monitor MA4)
Certain data on invisible earnings and payments.
Annual, 1988 edition £16.50

Census of Overseas Assets (Business Monitor MO4)
carried out every three years
1987 edition £10.50

Transport

Transport Statistics Great Britain, 1979-89
Annual £19.95

International Comparisons of Transport Statistics
1970-87 [5]
Occasional, £6.00

Quarterly Transport Statistics [2]
Quarterly £5.00. Annual £12.00

National Road Maintenance Condition Survey Report,
1989 [2]
Annual £5.00

National Road Maintenance Condition Survey
Sub-National Results 1989
Annual £2.50

Local road maintenance expenditure in England and
Wales 1988/1989
Annual £9.50

Motor Vehicle Registrations (Business Monitor MM1)
Monthly. Annual subscription £29.50

Traffic speeds in London; Outer Area Survey [2]
Occasional, price on application

Traffic speeds on London Roads; Inner Area Survey [2]
Occasional, price on application

Transport statistics for London
Annual £9.90

Motorway and Trunk Road Maintenance Expenditure in
England 1975/76-1987/88
Annual £6.80

Local Road Maintenance Expenditure in England
1982/83-1988/89
Occasional, price £6.30

---

1 Obtainable from address on page 110.
2 Department of Transport, Publication Sales Unit,
Building 1, Victoria Road, South Ruislip,
Middlesex HA4 0NZ.
3 From Scottish Office Library (address on page
109).

London Traffic Monitoring Report 1989
Annual £9.70

The Allocation of Road Track Costs 1990/91 United
Kingdom
Occasional, £5.00

Traffic in Great Britain [1]
Quarterly £8.00. Annual £27.00

Road Accidents, Great Britain, 1989, The Casualty
Report
Annual £9.75

Road Accidents Statistics English Regions, 1988
Annual £8.40

Road Casualties Great Britain [1]
Quarterly £4.00

Transport of Goods by Road in Great Britain 1988
Annual £9.95

Heavy Goods Vehicles in Great Britain 1989
Annual £7.50

International Road Haulage by UK registered vehicles
— Report on 1989
Annual, price on application

International Passenger Transport
Annual £14.50

Road Accidents in Scotland, 1989 [2]
Annual £5.00

Road Accidents in Wales No. 10 1989 [3]
Annual £3.00

Welsh Transport Statistics No. 6 1990 [3]
Annual £6.00

Scottish Transport Statistics, No. 10 1988 [2]
Annual £5.00

Statistical bulletins on various aspects of roads and
transport, Scotland [2] (eg 'Drunkenness and injury road
accidents in Scotland')
£1.25

Road Lengths in Great Britain 1988-1989
Annual £9.40

Road Goods Vehicles on Roll-on Roll-off Ferries to
Mainland Europe [1]
Quarterly £5.00

Survey of Heavy Goods Vehicles entering Britain via
Dover June 1988
Occasional £2.50

Goods vehicles in special types tax-class 1988
Annual £7.50

United Kingdom shipping industry — international
revenue and expenditure 1989
Annual £7.70

Bus and Coach statistics Great Britain 1988/89
Annual £8.70

Annual Vehicle Census, December 1989 [1]
Annual £3.00

International Comparisons of Transport Statistics
1970-1987
Part 3: Road vehicles, traffic, fuel and expenditure
Occasional £16.20

Seaborne Trade Statistics of the United Kingdom 1989
Annual £12.70

Merchant Fleet Statistics 1989
Annual £13.80

General Trends in Shipping [1]
Annual £12.00

Port Statistics 1989 [4]
Annual, price on application

Waterborne Freight in the United Kingdom, 1989[5]

Annual, price on application

Casualties to Vessels and Accidents to Men 1988
Annual £6.80

Vehicle Excise Duty Evasion in Great Britain 1989/90
Occasional £6.50

Vehicle Excise Duty Evasion in Great Britain
— a report on the method and data for reviewing the
timing of roadside surveys
Occasional £3.50

Road Lengths in Great Britain 1988-1989
Annual £9.40

Statistics of Road Lengths in Wales [3]
Occasional £2.00

Estimators for the National Road Traffic Series [1]
Occasional £3.00

Transhipment of UK Deep-sea Trade 1976-84 [4]
£25.00

Traffic monitoring in the Eastern Region [1]
Occasional £13.75

Bus and Coach Speed Survey: August and November
1986 [1]
Annual £4.00

Bus and Coach Statistics Great Britain 1989/90
Annual, price on application

London traffic monitoring report for 1987 [1]
Annual £12.00

National Road Maintenance Condition Survey, sub-
national results [1]
£1.50

Goods vehicles in restricted tax classes 1986 [1]
Occasional £2.50

Survey of heavy goods vehicles entering Britain via
Dover, Nov 1987 [1]
Occasional £2.50

---

1   Department of Transport, Publication Sales Unit,
    Building 1, Victoria Road, South Ruislip,
    Middlesex HA4 0NZ.
2   From Scottish Library (address on page 107).
3   Obtainable from E & SS Division, Welsh Office,
    Cathays Park, Cardiff CF1 3NQ.
4   From British Ports Federation, Victoria House,
    Victoria Place, London WC1B 4LL.
5   From Maritime and Distribution Systems, 28 City
    Road, Chester CH1 3AE.

Analysis of the International Earnings and Expenditure of the UK Shipping Industry 1986 [1]
Occasional £30.00

Vehicle Census, December 1987 [1]
Annual £3.00

International comparisons of Transport Statistics 1970-1987 [1]
Occasional £6.00

Overseas Travel and Tourism (Business Monitor MA6 — annual, and MQ6 — quarterly)
Includes the main results of the International Passenger Survey. Data on UK visitors overseas and visitors to the UK.
Annual, 1988 edition £6.50 Quarterly. £18.00 per annum, including postage.

UK Airlines — annual operating, traffic and financial statistics, 1989 (CAP 568) [2]
£11.00

UK Airports — annual statement of movements, passengers and cargo, 1989 (CAP 566) [2]
£11.00

UK Airlines — Monthly Operating and Traffic Statistics
Annual subscription (12 issues) £37.00. (Airmail £50.00.) Individual copies, £3.70 each

UK Airports: Monthly Statements of Movements, Passengers and Cargo
Annual subscription (12 issues) £37.00. (Airmail £50.00). Individual copies, £3.70 each

Reportable accidents to UK registered aircraft and to Foreign registered aircraft in UK Airspace 1988 (CAP 564)
£7.00

UK Air Miss Statistics — UK air misses involving commercial air transport [2]
(published 3 times a year) £5.00

Annual Punctuality Statistics — Heathrow, Gatwick, Manchester, Birmingham, Luton, and Stansted, April 1989 to March 1990: summary analysis. (CAP 571)
Annual £8.00

Annual Punctuality Statistics — Heathrow, Gatwick, Manchester, Birmingham, Luton and Stansted, April 1989 to March 1990: full analysis (CAP 572)
Monthly summary analysis £5.00, Full analysis £25.00

Accidents to UK aircraft and to large jet and turbo-prop transport aircraft worldwide in 1989. (CAP 569)
£2.50

Distribution and other Services

Retailing (Business Monitor SDA25)
Detailed statistics on types and numbers of businesses, turnover, gross margins, numbers employed etc.
Annual, 1987 edition £12.95

Retail Trade (Business Monitor SDM28)
Summary figures for retail trade and detailed index numbers of sales in various kinds of shops.
Monthly, £29.50 per annum including postage

Wholesaling (Business Monitor SDA26)
Annual, 1988 edition £12.70

Motor Trades (Business Monitor SDA27)
Annual, 1988 edition £14.20

Catering and allied trades (Business Monitor SDA28)
Annual, 1988 edition £16.20

Service Trades (Business Monitor SDA29)
Annual, 1988 edition £19.95

Computer Services (Business Monitor SDQ9)
Data on bureau services, software, hardware, consultations, training and employment.
Quarterly, £18.00 per annum including postage

Society

Manpower, earnings, retail prices etc.

Employment Gazette
Includes articles, tables and charts on manpower, employment, unemployment, hours worked, earnings, labour costs, retail prices, stoppages due to disputes etc.
Monthly £4.15 (annual subscription £43.50 including postage)

New Earnings Survey
Relates to earnings of employees by industry, occupation, region, etc., at April each year.
Six parts between September and December. Each part costs £9.75, or £56.00 including postage for all six parts

Time Rates of Wages and Hours of Work [3]
Loose-leaf publication with monthly updates.
Annual subscription £43.00

Labour Force Survey, 1973, 1975 and 1977; 1979; 1981; 1983 and 1984; 1985; 1986
Series of reports on the EC Labour Force Survey which collects data on various aspects of economic activity.
Price on application

Labour Market and Skill Trends 1991/1992 [4]
Describes relevant labour market and skill trends, and raises issues which those involved in planning training and enterprise will want to consider.

Labour Market Quarterly Report [4]
Provides a commentary, including tables and charts on current labour market trends and implications for training - employment, unemployment, and includes special features on specific labour market topics.
Quarterly free

England and Wales Youth Cohort Study [4]
Series of reports on education, training and employment histories of young people.

Training Statistics 1990
A reference volume containing statistical tables and charts compiled from a wide range of sources. Includes some technical commentary and a description of sources.
Annual £10.50

Retail Prices Indices 1914-1990
£10.95

National Online Manpower Information System (NOMIS) [4]
A nationally networked information system offering rapid access and integrated analysis for data on employment, unemployment, job vacancies, population and migration, at various geographical levels.

1	Department of Transport, Publication Sales Unit, Building 1, Victoria Road, South Ruislip, Middlesex HA4 0NZ.
2	Civil Aviation Authority, Printing and Publication Services, Greville House, 37 Gratton Road, Cheltenham, Gloucestershire GL50 2BN.
3	From Department of Employment, (HQ Stats A1), Watford WD1 7HH.
4	From Department of Employment, Moorfoot, Sheffield S1 4PQ (see page 36).

Civil Service Statistics 1989-90
Facts and figures about Civil Servants in 1988-89 and in previous years, including information on numbers by department, grade and location.
Annual, price on application

University Graduates 1989/90 [1]

Polytechnic First Degree and Higher Diploma Students 1989/90 [1]

Colleges and Institutes of Higher Education First Degree and Higher Diploma Students 1989/90 [1]

First Degree and Higher National Diploma Students from Centrally Funded Institutions in Scotland 1989/90.
Summaries of first destination and employment.
Annual £2.40 each

CSU Statistical Quarterly [1]
Analyses movements and changes within the graduate employment market and incorporates a salary survey of current rates for new and recent graduates.
Annual subscription £25.00. £8.00 Quarterly

The Supply of Polytechnic Graduates 1990–1992: Trends and Predictions [1]
Gives estimates of the numbers graduating and of the numbers available for employment over the next three years.
Annual £9.00 each

Output of UK Universities by Institution and Discipline 1990 [1]

Output of Polytechnics by Institution and Discipline 1990 [1]

Output of Colleges of Higher Education by Institution and Discipline. Actual output of universities, polytechnics and colleges by institute, subject, sex, and domicile.
Annual £8.25 each

Taxation

Inland Revenue Statistics 1989
Contains statistics related to all taxes administered by the Inland Revenue, ie income tax, corporation tax, petroleum revenue tax, inheritance tax, capital gains tax and stamp duties. Also contains some tables on numbers of taxpayers and analyses of taxable incomes and income tax by range and source of income. Includes data on rateable values, numbers and values of conveyances of land and buildings, agricultural land transactions, estimates of the distribution of personal wealth, as well as data on the Business Expansion Scheme, employee share schemes profit-related pay, fringe benefits and mortgage interest relief.
Annual £15.50

Report of the Commissioners of HM Inland Revenue
Includes statistics of all taxes administered by the Inland Revenue; in particular tax collected, tax recovered as a result of investigation activities, tax remitted or written off, and costs of collection.
Annual £13.70

Report of the Commissioners of HM Customs and Excise
Statistics related to all taxes administered by Customs and Excise including the duties on tobacco, alcoholic drink, hydrocarbon oil, betting and gaming, customs duties, agricultural levies, matches and mechanical lighters, car tax and Value Added Tax.
Annual, price on application

Betting and Gaming Bulletin [2]
Revenue statistics.
Monthly. Annual subscription £18.00 including postage

Standard of living
(See also Retail prices etc., page 99, and note under Agriculture and Fisheries page 96; for consumers' expenditure consult *UK National Accounts*, page 94.)

Household Food Consumption and Expenditure (National Food Survey)
Report on food consumption, expenditure and nutrition by type of household. Latest report for 1984. Unpublished data and analyses may also be purchased.
Annual £13.00

Population and households
Statistics on population and households are collected in the periodic censuses conducted by the OPCS and the Registrars General for Scotland and Northern Ireland. The most recent census was taken in April 1991.
Statistics from the 1981 Census are available in reports for each county in England and Wales (Regions and Island Areas in Scotland), and in a national and regional report. The same range of statistics is available for small areas (Small Area Statistics) throughout Britain. There are also summary reports of key statistics for local authorities, towns and cities, parliamentary constituencies and wards.
Full details of these and other census publications can be obtained from the address on page 108.

There is a series of national reports on particular topics including family statistics (births, marriage, divorce), mortality, morbidity (including cancer, infectious diseases and congenital malformations), abortion, international migration; vital statistics (population, births, deaths and internal migration) and electoral statistics (based on electoral rolls).
For annual mid-year estimates of population and data to local authority level consult:

Key Population and vital statistics, 1989
Annual £9.10

Population Estimates, Scotland, 1989
£2.00

The Registrar General's Annual Report for Northern Ireland
Price on application

The Registrar General's Annual Report for Scotland [3]
Price on application

---

1 From Central Services Unit, Crawford House, Precinct Centre, Manchester M13 9EP.
2 From Business and Trade Statistics Limited, Lancaster House, More Lane, Esher, Surrey KT10 8AP, Telephone 0372 63121.
3 From General Register Office (Scotland), Population Statistics Branch, Ladywell House, Edinburgh EH12 7TF.

Population Trends
Contains regular series of tables on subjects for which OPCS is responsible in England and Wales, in addition to articles on a variety of population and medical topics. For similar information for Scotland and Northern Ireland consult the publications of the appropriate Registrar General.
Quarterly, various prices

Population Projections
(i) Projections for the UK and selected constituent countries over the next 40 years
Latest projections, published on microfiche, based on mid-1987 population.
Biennial £10.50

(ii) Projections for the English regions, counties, metropolitan districts and London boroughs for 1985, 1991 and 2001
Latest projections based on provisional mid-1985 population estimates.
Biennial £7.40

(iii) 1987-based Population projections for the counties of Wales [1]
Biennial £3.00

(iv) Estimates of Numbers of Households in England, the regions, counties, metropolitan districts and London boroughs to 2001 [2]
Periodic

(v) Projections for standard areas of Scotland over the next 10 to 15 years [3]
Latest projections based on the mid-1989 population estimates.
Biennial summary
Also available on floppy disk, prices on application

(vi) 1987-based household projections for the counties of Wales [1]
Biennial £10.00

OPCS Monitors
Available on subscription or by standing order from Information Branch (Dept GS), Office of Population Censuses and Surveys, and designed for the quick release of figures within the range of OPCS activities. These include the most recent population estimates and projection figures.
Rates available on application

Family Expenditure Survey
Shows, in great detail, income and expenditure by type of household for the UK and includes some regional analyses. Latest report for 1989.
Annual £18.00

General Household Survey
A continuous sample survey of households relating to a wide range of social and socio-economic policy areas. Latest report for 1988.
Annual £16.50

Education
(See also Manpower, etc., page 99) Statistical bulletins [4] (free) issued regularly, covering a variety of educational topics.

Education Statistics for the UK, 1990 Edition

Statistics of Education [5]
Six sets of statistical volumes about education are available. They refer to – schools; school leavers and examinations; further education; teachers in service; finance and awards; further education student/staff ratios.
Annual £12.00 per set (individual tables 25p, minimum charge £1.50 for 6 pages or less)

Education Facts and Figures (England) [6]
A free Facts Card containing a small selection of up-to-date basic education statistics.
Annual

Selected National Education Systems [8]
A free volume containing a description of the educational systems of six countries (France, Italy, USA, Japan, Germany and the Netherlands) as an aid to international comparisons.

Selected National Education Systems II [8]
A second free volume containing a description of six further countries (Australia, Belgium, Canada, Denmark, Spain and Sweden).

Statistics of Education in Wales: Schools No. 4 1990 [1]
Annual £4.00

Statistics of Education in Wales: Higher and Further Education No. 3 1990 [1]
Annual, price on application

Basic Educational Statistics Fact Card, Scotland [7]
Biennial, free

Statistical bulletins covering education in Scotland [2] (eg 'Pupils and teachers in education authority primary and secondary schools', 'School leavers qualifications', 'Higher education projections')
Occasional £1.25

University Statistics series [9]
Volume 1 – Students and Staff
Volume 2 – First Destinations of University Graduates
Volume 3 – Finance
Annual, prices on application

University Management Statistics and Performance Indicators in the UK
Annual, price on application

---

1 Obtainable from E & SS Division, Welsh Office, Cathays Park, Cardiff CF1 3NQ.

2 Further details from Department of the Environment, HDS12 Division, Room N2/14, 2 Marsham Street, London SW1P 3EB.

3 From General Register Office (Scotland), Population Statistics Branch, Ladywell House, Edinburgh EH12 7TF.

4 From Department of Education and Science, Scottish Education Department and Department of Education for Northern Ireland (addresses on pages 106, 109 and 110).

5 From Department of Education and Science, Room 0100, Mowden Hall, Staindrop Road, Darlington DL3 9BG.

6 From Department of Education and Science (address on page 106).

7 From Scottish Office Library (address on page 107).

8 From Department of Education and Science, Room 304, Mowden Hall, Staindrop Road, Darlington DL3 9BG.

9 From Universities Statistical Record, PO Box 130, Cheltenham, Gloucestershire GL50 1SE. Telephone 0242 225902.

Home affairs

In addition to detailed annual publications, available either from HMSO or the Home Office, the Home Office issues a series of Statistical Bulletins, some regular (quarterly or annual) and some on an *ad hoc* basis. A comprehensive list of bulletins issued in the last year is available from the Home Office Statistical Department (address on page 107).

Betting Licensing Statistics, Great Britain, 1988/9 (Bulletin Issue 32/90) [1]
Annual £2.50

Betting Licensing Statistics, Great Britain, Supplementary Tables, 1989/90 [1]
Annual £40.00

Statistics of Scientific Procedures on Living Animals, Great Britain, 1989 (Cm 743)
Annual £8.50

Fire Statistics, United Kingdom 1988 [1]
Annual £5.50

Control of Immigration: Statistics, United Kingdom, 1989 (Cm 726)
Annual £14.80

Control of Immigration: Statistics, (Bulletin) [1]
Biennial £2.50

Immigration from the Indian Sub-continent, 1988 (Bulletin Issue 44/89) [1]
Annual £2.50
(For migration, see also Population and households, page 100.)

Refugee Statistics, United Kingdom, 1989 (Bulletin Issue 22/90) [1]
Annual £2.50

Citizenship Statistics, United Kingdom, 1989 (Bulletin Issue 11/90) [1]
Annual £2.50

Liquor Licensing Statistics, England and Wales, 1988/9 [1]
Triennial £2.50

Liquor Licensing Statistics, England and Wales; Supplementary Tables 1988/9 [1]
Triennial £40.00

Local Government Elections, England and Wales, 1989 (Bulletin Issue 35/89) [1]
Annual £2.50

Election Expenses, June 1987, House of Commons Paper (HCP 426 (87/88))
Published after each general election £11.40

European Parliamentary Election Expenses, United Kingdom, June 1989 [1]
Published after each election £2.50

Statistics of deaths reported to Coroners, England and Wales, 1989 (Bulletin Issue 14/90) [1]
Annual £1.50

Domestic proceedings in magistrates' courts, England and Wales, 1987 (Bulletin Issue 21/90) [1]
Annual £2.50

Justice and law

In addition to detailed annual publications, available either from HMSO or the Home Office, the Home Office issues a series of Statistical Bulletins, some regular (quarterly or annual) and some on an *ad hoc* basis. A comprehensive list of bulletins issued in the last year is available from the Home Office Statistical Department (address on page 107).

Criminal Statistics, England and Wales, 1989 (Cm 1322)
Annual £17.60

Supplementary Tables 1989, annual
Vol. 1 Proceedings in magistrates' courts [1]
£16.00
Vol. 2 Proceedings in the Crown Court
£16.00
Vol. 3 Police force areas and court areas: recorded offences, firearms offences, court proceedings, cautions
£14.00
Vol. 4 Convictions, cautions, specific aspects of sentencing, appeals, prerogative of mercy
£8.00
Vol. 5 Court proceedings by Petty Sessional Division and Commission of the Peace Area
£12.00

Judicial Statistics, England and Wales, 1989 (Cm 745)
Annual £10.60

Information Bulletin: Criminal Legal Aid 1989 [2]
Free

Statistical bulletins covering aspects of the criminal justice system in Scotland [3] (eg 'Homicide in Scotland, 1983–87', (Bulletin Issue 5/89). 'Recorded crime in Scotland 1989', (Bulletin Issue 2/90). 'Criminal Proceedings in Scottish Courts' (Bulletin Issue 1/90).
Occasional £1.25

Prison Statistics, England and Wales, 1989 (Cm 1221)
Annual £17.40

Statistics of offences against prison discipline and punishments, England and Wales, 1989 (Cm 1236)
Annual £7.95

Prisons in Scotland 1988/89 Report, Annual Command Paper
Price on application

Statistical bulletins about prisons in Scotland [3] (eg 'Prison statistics, Scotland 1988' (Bulletin Issue 6/89))
Occasional £1.25

Statistics of mentally disordered offenders 1987 and 1988 (Bulletin Issue 16/90) [1]
£2.50

The prison population 1989 (Bulletin Issue 12/90) [1]
Annual £2.50

Updated projections of long-term trends in the prison population to 1998 (Bulletin Issue 33/90) [1]
Annual £2.50

---

1 From Home Office Statistical Department (address on page 107).
2 From Lord Chancellor's Department (address on page 107).
3 From Scottish Office Library (address on page 109).

Probation Statistics, England and Wales, 1989 [1]
Annual £5.50

Summary Probation Statistics, England and Wales, 1989 (Bulletin Issue 28/90) [1]
Annual £2.50

Offences relating to motor vehicles, England and Wales, 1989 (Bulletin Issue 34/90) [1]
Annual £2.50
(Supplementary Tables £4.00)

Offences of Drunkenness, England and Wales, 1988 (Bulletin Issue 40/89) [1]
Annual £2.50

Statistics of drug addicts notified to the Home Office, United Kingdom, 1989 (Bulletin Issue 7/90) [1]
Annual £2.50

Statistics of drug addicts notified to the Home Office, United Kingdom, 1989, Area Tables [1]
Annual £2.50

Statistics of the misuse of drugs: Seizures and offenders dealt with, United Kingdom, 1989 (Bulletin Issue 24/90) [1]
Annual £2.50

Statistics of the misuse of drugs: Seizures and offenders dealt with, United Kingdom, 1989, Supplementary and Area Tables [1]
Annual £2.50 each

Notifiable offences recorded by the police, England and Wales (Bulletin) [1]
Quarterly £1.50

Statistics on the operation of the prevention of terrorism legislation (Bulletin) [1]
Quarterly £2.50

Crime statistics for the Metropolitan Police Districts analysed by ethnic group (Bulletin Issue 5/89) [1]
Ad hoc £2.50

Statistics of breath tests, 1989 (Bulletin Issue 25/90) [1]
Annual £2.50

Statistics on the operation of certain police powers under the Police and Criminal Evidence Act (Bulletin) [1]
Quarterly £2.50

Statistics of the time taken to process criminal cases in magistrates' courts (Bulletin) [1]
Three times a year £2.50

Reconvictions and recalls of Life Licensees and mentally disordered offenders (Bulletin Issue 27/90) [1]
Annual £2.50

Health, personal social services, safety and social security

Health and Safety Statistics 1988-89
Published as a supplement to the November 1990 issue of the Dept. of Employment's periodical *Employment Gazette*
Monthly £4.15. Annual subscription £43.50 including postage

Health and Safety Commission Annual Report 1989–90
Annual £10.00

Accidents in Service Industries: 1988/89 Health and Safety Statistics for premises inspected by Local Authorities
Annual, free

Health and Personal Social Services Statistics for England 1990
Annual £10.85

Health and Personal Social Services Statistics for Wales No. 17 1990 [2]
Annual £5.00

Various publications on aspects of hospital activity and social services in Wales [2]
Annual or twice yearly £3.00

Social Security Statistics
Annual £15.90

Abstract of Statistics for Index of Retail Prices. Average Earnings, Social Security Benefits and Contributions [3]
Annual £15.00

Statistical bulletins on various aspects of Health and Personal Social Services activity in England [4]
£2.00

Scottish Health Statistics, 1990 [5]
Annual £5.95

Statistical publications on aspects of health services and health in Scotland [5]
Annual £1.00 to £10.00

Statistical bulletins on various aspects of social services in Scotland [5] (eg 'Referrals of children to reporters and children's hearings 1989' *£1.25*, 'Children in care or under supervision 1989' *£1.25*, 'Staff of Scottish Social Work Departments 1989' *£1.25*, 'Residential accommodation 1989' *£1.00*, 'Home care services, day care establishments and day care services 1988(2)' *£1.00*, 'Community Service by offenders 1989' *£1.25.*)

Hospital In-Patient Enquiry, Main Tables, 1985
Annual £10.50 + VAT

Hospital In-Patient Enquiry, Summary Tables, 1985
£6.20 + VAT

Hospital In-Patient Enquiry, Trends 1979-1985
£10.00 + VAT

Hospital In-Patient Enquiry, Main Tables, 1984
£9.50 + VAT

Hospital In-Patient Enquiry, Summary Tables, 1984
£6.20 + VAT

---

1 From Home Office Statistical Department, (address on page 107).
2 Obtainable from E & SS Welsh Office, Cathays Park, Cardiff CF1 3NQ.
3 From HQSR8A, Room A2215, DSS, Central Office, Newcastle upon Tyne NE98 1YX.
4 From DSS, Information Division, Canons Park, Government Buildings, Honeypot Lane, Stanmore, Middlesex HA7 1AY.
5 From Scottish Health Service (address on page 107).

Hospital Statistics, Form SH3 Regional and National Summaries for 1986

Also see relevant columns of the annual reports of the Registrars General for Scotland and N Ireland.

Mental Health Statistics for Wales No. 10 1990 [1]
Annual £5.00

For mortality statistics and morbidity, including cancer and infectious diseases, see Population and households, page 100.

Activities of Social Services Departments: Year ended 31 March 1989
Annual £3.00

Welsh Hospital Waiting List Bulletin 1990 No. 2
Staff of Social Services Departments: Year ended 30 September 1990
Annual £3.00 or twice yearly £3.00

Children in Care or under supervision orders in Wales: Year ended 31 March 1989
Annual £3.00

Residential Accommodation for the elderly, blind and physically disabled: Year ended 31 March 1990
Annual £3.00

Key Statistical Indicators for NHS Management in Wales No. 9 1990
Annual £3.00

Environment

Digest of Environmental Protection and Water Statistics, No. 12, 1989
Annual £10.30

Statistical Bulletins (88) 1-4 [2] (Supplement to the above Digest)
Annual £3.00 each

Environmental Digest for Wales No. 5 1990 [1]
Annual £6.00

Scottish Environmental Statistics No. 2 1989 [3]
Biennial £5.00

Statistical bulletins on various aspects of the environment in Scotland [3] (eg 'Radioactive waste disposals from nuclear sites in Scotland, 1984–88', 'Chernobyl incident: monitoring for radioactivity in Scotland', 'Land use change statistics' 1987 & 1988)
£1.25

Survey of Derelict Land in England, 1982 [2]
Occasional £5.90

Survey of land for mineral working in England, 1982 [2]
Occasional £15.50

For production, consumption and trade statistics on UK minerals, see page 96.

Land use change in England, Statistical Bulletin (90)5 data recorded during 1989 [2]
Periodic £3.00

Development Control Statistics: England 1988/1989 [2]
Periodic £4.75

Quarterly press releases on planning applications and decisions [4]
Available on application

Land Registers [5]
The DOE's Computerised Land Register contains information about 8,300 sites (86,500 acres) of unused or under-used land in the public sector. Public access to the Land Register is available through the Department and most regional offices.
Initial charge £27.00

Digitised Boundaries [6]
1981 Census wards, districts, counties including policy areas, National Parks, AONB, Green Belt etc — England and Wales. For use with geographic information systems.
Available on magnetic tape through DOE.

Enterprise Zone Information 1987/88, December 1990
Annual £7.80

Monitoring Landscape Change, July 1986 [7]
Price on application

Overseas aid

British Aid Statistics [8]
Statistics of UK economic aid to developing countries.
Annual £6.75 (or £7.25 including postage)

---

1 Obtainable from E & SS Division, Welsh Office, Cathays Park, Cardiff CF1 3NQ.
2 From Department of the Environment, Publications Sales Unit, Victoria Road, South Ruislip, Middlesex HA4 0NZ.
3 From Scottish Office Library (address on page 109).
4 From LGS Division, Department of the Environment, Room P1/177B, 2 Marsham Street, London SW1P 3EB.
5 Further information is available from UDCD Division, Room P2/081, Department of the Environment, 2 Marsham Street, London SW1P 3EB.
6 Further information is available from MAP Library, OSD2 Division, Department of the Environment, Room P2/162, 2 Marsham Street, London SW1P 3EB.
7 Available from Hunting Technical Services Ltd, Elstree Way, Borehamwood, Herts WD6 1SB.
8 From the Library, Overseas Development Administration (address on page 108).

DEPARTMENTAL RESPONSIBILITIES AND CONTACT POINTS

Central Statistical Office

Great George Street, London SW1P 3AQ	**071-270** (followed by extension number)
National accounts, balance of payments, index of output of the production industries, cyclical indicators, financial statistics, input-output statistics, development of social statistics, regional statistics	exts. 6363/6364
Macro-Economic Databank	ext. 6386
Manpower qualified in science, engineering and technology. Research and Development (R&D) expenditure and employment, Cabinet Office Annual Review of Government Funded R&D	ext. 6068

Central Statistical Office, Cardiff Road, Newport, Gwent NP9 1XG	**0633 81** (followed by extension number)
General enquiries	ext. 2973
Producer price indices	exts. 2106/2584
Price index numbers for current cost accounting	ext. 2173
Production statistics:	
Annual, quarterly and monthly production enquiries (specific products)	ext. 2973
Annual census of production statistics — manufacturing	ext. 2455
— construction	ext. 2435
— minerals raised enquiry	ext. 2082
Enquiries into purchases	ext. 2893
Index of industrial production	ext. 2786
Distribution and service statistics:	
Retail enquiry	ext. 2710
Wholesaling and dealing enquiry	ext. 2266
Catering enquiry	ext. 2266
Property enquiry	ext. 2710
Service trades enquiry	ext. 2264
Motor trades enquiry	ext. 2180
Monthly retail sales index	ext. 2987
Other short-period enquiries	ext. 2609
Capital expenditure and stocks and investments in:	
Manufacturing — stocks	ext. 2213
— capital expenditure and investment intentions	ext. 2215
Distributive and service industries	ext. 2215
Analysis of company accounts and rate of returns on capital	ext. 2580
UK Directory of manufacturing businesses	ext. 2991

Millbank Tower, London SW1P 4QU	**071-217** (followed by extension number)
Family Expenditure Survey	exts. 4245/4255
Insurance companies and pension funds	ext. 4205
Personal Pensions	ext. 4203
Credit business and consumer credit	ext. 4342
Overseas direct investment, overseas transactions in insurance, overseas takeovers, and civil aircraft expenditure	exts. 4333/4238
Overseas transactions: films and television enquiry, royalties	exts 4386/4728
Acquisitions and mergers	ext. 4202
Company liquidity and rates of return on capital	ext. 4334
Survey Control Unit enquiries only	ext. 4320

Exchange House, 60 Exchange Road, Watford, Herts WD1 7HH	**092381** (followed by extension number)
Index of retail prices	ext. 5377

Other government departments

Ministry of Agriculture, Fisheries and Food	**Whitehall Place, West, London SW1A 2HH**	**071-270** (followed by extension number)
	Farm incomes	exts. 8618/8621
	Agricultural and food statistics for overseas countries	ext. 8632
	Food Industry statistics	ext. 8559
	National food survey	ext. 8562
	Agricultural wages and employment	ext. 8646
	Whitehall Place East, London SW1A 2HH	**071-270** (followed by extension number)
	Agricultural rents, land prices etc.	ext. 8371
	Ergon House, c/o Nobel House, 17 Smith Square, London SW1P 3HX	**071-238** (followed by extension number)
	Commodity production statistics	ext. 6402
	Fisheries	ext. 5913
	Economic accounts for agriculture	ext. 6402
	Government Buildings, Epsom Road, Guildford GU1 2LD	**0483 68121**
	Agricultural censuses and surveys	ext. 3520
	Agricultural price indices	ext. 3712
	Producer price index (manufactured food and soft drinks)	ext. 3712
	Market prices (horticulture and livestock)	ext. 3499
HM Customs and Excise	**New King's Beam House, 22 Upper Ground, London SE1 9PJ**	**071-620 1313**
	Statistics related to all indirect taxes administered by the department, other than VAT and car tax	071-382 5041
	VAT and car tax	071-382 5036
	Monthly trade statistics and importer details	071-382 5046
	are available from marketing agents appointed by HM Customs and Excise. A list of names and addresses is available from the following address: HM Customs and Excise, Tariff and Statistical Office, Unit 51, Room 716, Portcullis House, 27, Victoria Avenue, Southend-on-Sea, Essex SS2 6AL	0702 367485
Ministry of Defence	**GF Stats 2, Northumberland House, Northumberland Avenue, London WC2N 5BP**	**071-218 0339**
	Volume II of the Annual Statement on the Defence Estimates and enquiries on published defence statistics	
Department of Education and Science	**Elizabeth House, York Road, London SE1 7PH**	**071-934** (followed by extension number)
	Statistics of schools, school-leavers, teachers, further and higher education, educational finance and awards and further education student/staff ratios.	
	International comparisons	exts. 9038/9037
Department of Employment	**Headquarters Buildings, East Lane, Halton, Runcorn WA7 2DN**	**0928 715151**
	Employment and hours	ext. 2555
	Exchange House, 60 Exchange Road, Watford, Herts WD1 7HH	**092381** (followed by extension number)
	Employment census	ext. 5312
	Manual workers' earnings	ext. 5222
	Wage rates, basic hours	ext. 5219
	Index of average earnings	exts. 5208/5214
	New earnings survey	ext. 5234
	Stoppages of work due to industrial disputes	exts. 5266/5265
	Employers' labour costs	ext. 5219
	Caxton House, Tothill Street, London SW1H 9NF	**071-273** (followed by extension number)
	Labour force survey	ext. 5586
	Unemployment and vacancies	ext. 5532
	Redundancy statistics	ext. 5530
	Tourism statistics	ext. 5507
	Unit wage costs and productivity	ext. 5535
	Public enquiry office	ext. 6969
	Moorfoot, Sheffield S1 4PQ	**0742-59** (followed by extension number)
	Labour Market Quarterly Report	ext. 4075
	NOMIS	ext. 4086
	Labour Market and Skill Trends 1991/92	ext. 4075
	Youth Cohorts Study	ext. 4194

Department of Energy	1 Palace Street, London SW1E 5HE Fuel and energy statistics	071-238 (followed by extension number) exts. 3606/3576
Department of the Environment	2 Marsham Street, London SW1P 3EB	071-276 (followed by extension number)
	Building materials	ext. 4761
	Construction indices	ext. 3460
	Construction output, orders and employment	ext. 3526
	Council house sales	ext. 3505
	Digitised boundaries	ext. 4539
	Enterprise Zones	ext. 4468
	Environmental protection	ext. 8422
	Household projections	ext. 4192
	House prices	ext. 3497
	Housing (general)	ext. 3496
	Inner Cities	ext. 4429
	Landscape change	ext. 8421
	Land Registers	ext. 4463
	Land use change statistics	ext. 4533
	Local government financial statistics	ext. 3033
	Planning and land use	ext. 4168
	Population projections	ext. 4524
	Survey of derelict land in England	ext. 4435
	Survey of land for mineral working in England	ext. 3963
Government Actuary's Department	22 Kingsway, London WC2B 6LE Surveys of occupational pensions schemes; national population projections (alternative contact point to OPCS)	071-242 6828
Health and Safety Executive	Daniel House, Stanley Road, Bootle, Merseyside L20 3QZ Health and safety statistics	051-951 4000
Department of Health	Friars House, 157-168 Blackfriars Road, London SE1 8EU	071-972 2000
	Financial statistics	ext. 23128
	Hannibal House, Elephant and Castle, London SE1 6TE	071-972 2000
	Statistics of health services:	
	Manpower	ext. 22350
	Non-psychiatric hospital services	ext. 22193
	Mental illness and mental handicap services	ext. 22210
	Community and environmental health services	ext. 22201
	Statistics of personal social services	ext. 22213
	14 Russell Square, London WC1B 5EP	071-636 6811
	Family Health services	ext. 3070
Home Office	Lunar House, 40 Wellesley Road, Croydon, Surrey CR0 9YD Statistics on the administration of justice, criminal and penal matters, immigration and nationality, misuse of drugs, fire, betting and liquor licensing, community relations, scientific procedures on living animals, coroners and elections	081-760 2850
Board of Inland Revenue	Somerset House, Strand, London WC2R 1LB Statistics of direct taxes, corporate income, surveys of personal incomes, personal wealth and conveyancing	071-438 7370
Lord Chancellor's Department	Information Management Unit, 6th Floor, Trevelyan House, 30 Great Peter Street, London SW1P 2BY Judicial statistics relating to: The Appellate Courts, High Court, county courts and other civil courts The Crown Court, judges' sittings, courtroom utilisation and legal aid	071-210 8659
		exts. 2003/4

| Overseas Development Administration | The Library,
Abercrombie House,
Eaglesham Road, East Kilbride,
Glasgow G75 8EA | 03552-41199 |
| | Statistics of British aid to developing countries | ext. 3599 |

Office of Population Censuses and Surveys	St Catherines House, 10 Kingsway, London WC2B 6JP	071-242 0262
	Population and vital statistics (England & Wales):	
	Sub-national population projections (England)	ext. 2180
	International and internal migration	ext. 2182
	National population projections, marriages, divorces and general enquiries	exts. 2166/2168/2169
	Ethnic statistics	exts. 2302/2148
	Deaths, cancer, infectious diseases	exts. 2229/2184
	Abortions	exts. 2229/2184
	Births	exts. 2162/2165
	Social surveys	exts. 2287/2206
	Census of population — general enquiries	exts. 2008/2009
	Longitudinal study — general enquiries	ext. 2031
	Population estimates and projections	exts. 2541/2482/2150
		exts. 2235/2237
	The OPCS Reference Library at St Catherines House holds copies of all OPCS and General Register Office publications together with national and international publications on OPCS subjects. It also holds the Census Small Area Statistics on microfiche for areas throughout England and Wales. Visitors by appointment only.	
	Titchfield, Fareham, Hampshire PO15 5RR	0329 42511
	Supply of special and unpublished census statistics	ext. 3800
	National and local estimates of population	ext. 3265
	Note: Comparable information for Scotland and Northern Ireland is obtainable from:	
	General Register Office (Scotland), Ladywell House, Ladywell Road, Edinburgh EH12 7TF	031-334 0380
	Vital statistics	exts. 243/227
	Census Customer Services	exts. 254/266
	General Register Office (Northern Ireland), Oxford House, 49-55 Chichester Street, Belfast BT1 4HL	0232 235211 ext. 2341
	Census Office, 5A Frederick Street, Belfast BT1 2LW	0232 244711 exts. 205/395

| Department of Social Security | Central Office,
Newcastle upon Tyne NE98 1YX | |
| | Social security statistics | 091-279 7373 |

Department of Trade and Industry	Export Market Information Centre, the DTI's library for British exporters	071-215 5444/5445
	provides up-to-date and comprehensive trade statistics for all countries, and general statistical publications from all over the world. Foreign directories, development plans and mail order catalogues are also held. The reading room is open to the public Monday to Friday 9.30 am — 5.30 pm (last admission 5 pm). Copying facilities available.	
	151 Buckingham Palace Road, London SW1W 9SS	071-215 (followed by extension number)
	Insolvencies	exts. 1921/5602
	Overseas trade analysed in terms of industries	ext. 1907
	Manufacturing industry, import penetration and export sales ratio	ext. 1907

Minerals, metals, chemicals, plastics and pollution control equipment		ext. 1887
Consumer goods		ext. 1883
Textiles, natural and man-made fibres, tobacco, paper and board, printing and publishing		ext. 1883
Industrial and commercial refrigeration, ventilation and air conditioning		ext. 1883
Primary batteries and miscellaneous manufacturing industries		ext. 1883
Vehicles, agricultural machinery and tractors		ext. 1883
Engineering (including IT industries)		ext. 1882

Training Agency	**Moorfoot, Sheffield S1 4PQ**	**(see Department of Employment)**

Department of Transport	**2 Marsham Street, London SW1P 3EB and Romney House, 43 Marsham Street, London SW1P 3PY**	**071-276** (followed by extension number)
	Road accidents	ext. 8774
	Buses, coaches and taxis	ext. 8043
	Road conditions	ext. 8773
	Public expenditure on roads	ext. 8773
	Road goods transport	ext. 8188
	Road traffic and road lengths	ext. 8018
	National speed survey	ext. 8774
	National travel survey	ext. 8033
	New vehicle registrations	ext. 8699
	Ports statistics	ext. 8525
	Shipping statistics	ext. 8445
	Railway accidents	ext. 4876
	Railway passengers and freight	ext. 4838
	International transport statistics	ext. 8515
	International road haulage statistics	ext. 8518
	Domestic waterborne freight	ext. 8525

HM Treasury	**Parliament Street, London SW1P 3AG** General enquiries	**071-270 4860**

Scottish Office	**New St Andrew's House, Edinburgh** (postcodes as at subject headings)	**031-244** (followed by extension number)
	General enquiries (EH1 3SX)	ext. 4991
	Transport (EH1 3SX)	ext. 4981
	Planning/environment (EH1 3SX)	ext. 4990
	Local goverment financial statistics (EH1 3SX)	ext. 5257
	Scottish Office Library, Publication Sales, Room 2/65, New St Andrew's House, Edinburgh EH1 3TD	
	St Andrew's House, Edinburgh EH1 3DE	**031-244** (followed by extension number)
	Crime and prisons	ext. 2225
	Court proceedings	ext. 2227
	Housing	ext. 2687
	Scottish Education Department, 43 Jeffrey Street, Edinburgh EH1 1DH	**031-244** (followed by extension number)
	Education	ext. 5375
	Social work	ext. 5431
	Pentland House, 47 Robbs Loan, Edinburgh EH14 1UE	**031-244** (followed by extension number)
	Agriculture:	
	General statistics	ext. 6149
	Fisheries	ext. 6438

Scottish Office *(Cont'd)*	**Alhambra House, 45 Waterloo Street, Glasgow G5 6AS**	**041-248 2855**
	Scottish economic matters	ext. 5451
	Scottish Health Service, Common Services Agency, Information and Statistics Division, Trinity Park House, Edinburgh EH5 3SQ	**031-552 6255**
	Health statistics	ext. 2707
	Population and vital statistics, see under GRO Scotland, page 108	
Welsh Office	**Economic and Statistical Services Division, Crown Building, Cathays Park, Cardiff CF1 3NQ**	**0222 82** (followed by extension number)
	Economic statistics	ext. 5065
	Financial statistics	ext. 5079
	Demographic statistics	ext. 5085
	Education statistics	ext. 5057
	Health statistics	ext. 5080
	Housing statistics	ext. 5061
	Personal social services statistics	ext. 5041
	Agricultural statistics	ext. 5052
	Planning statistics	ext. 5062
	Transport statistics	ext. 5062
	General enquiries	ext. 5087
Northern Ireland Departments	**Stormont, Belfast BT4 3SW**	**0232 763210**
	Social statistics	ext. 2449
	Economic statistics	ext. 2061
	Prices	ext. 2061
	Housing statistics	ext. 2023
	Dundonald House, Upper Newtownards Road, Belfast BT4 3SF	**0232 650111**
	Agricultural statistics	ext. 785
	Netherleigh House, Massey Avenue, Belfast BT4 2JS	**0232 763244**
	Manpower statistics	ext. 2392
	Employment statistics	ext. 2474
	Production and trade statistics	ext. 2494
	Castle Buildings, Stormont Grounds, Belfast BT4 3SW	**0232 763939**
	Health and personal social services	ext. 2800
	Statistics of social security benefits	ext. 2062
	Dundonald House, Upper Newtownards Road, Belfast BT4 3SF	**0232 763255**
	Criminal and custodial statistics	ext. 431
	Oxford House, 49-55 Chichester Street, Belfast BT1 4HF	**0232 235211**
	Population and vital statistics	ext. 2341
	Rathgael House, Balloo Road, Bangor, Co. Down BT19 2PR	**0247 270077**
	Education statistics	ext. 2676
	Stormont, Belfast BT4 3SW	**0232 763210**
	Transport statistics	ext. 2528
	Road statistics	ext 2528

Remember:
If you have difficulty in finding the right contact, phone or write to the.
Library, Central Statistical Office, Government Buildings, Cardiff Road,
Newport, Gwent NP9 1XG
(Telephone 0633 812973; Fax: 0633 812599)
Or contact the Central Statistical Office, Great George Street, London SW1P
3AQ
(Telephone 071-270 6363/6364).

CENTRAL STATISTICAL OFFICE

Monthly Digest of Statistics

The Monthly Digest of Statistics provides basic information on 20 subjects including population, employment and prices, social services, production and output, energy, engineering, construction, transport, catering, national and overseas finance and the weather. It contains mostly runs of monthly and quarterly estimates for at least two years and annual figures for several more.

Price £6.50

Central Statistical Office publications are published by HMSO.
They are obtainable from HMSO bookshops and through booksellers.

Annual Abstract of Statistics 1991

Price £19.95

ISBN 0 11 620446 X

Central Statistical Office publications are published by HMSO.
They are obtainable from HMSO bookshops and through booksellers.

For about 140 years the *Annual Abstract of Statistics* has probably been the most quoted source of statistics about the United Kingdom.

348 tables in 18 separate chapters cover just about every aspect of economic, social and industrial life. Most of the data in the Abstract are annual and cover periods of about 10 years.

Printed in the United Kingdom for HMSO
Dd295358 9/91 C35 G3390 10170